Special Publications 72

OIL AND GAS EXPLORATION

Methods and Application

Edited by
Said Gaci
Olga Hachay

This Work is a co-publication between
the American Geophysical Union and John Wiley & Sons, Inc.

Library of Congress Cataloging-in-Publication Data is available.

ISBN: 978-1-119-22742-7

Cover image: *A set of Vibroseis trucks used to generate seismic signals in a seismic land acquisition. A Vibroseis truck used to provide a seismic source. The source signal is provided by a servo-controlled hydraulic vibrator or shaker unit mounted on a mobile base unit.*

Printed in the United States of America

10 9 8 7 6 5 4 3 2 1

CONTENTS

CONTRIBUTOR LIST

Mohand Amokrane Aitouch
Physics of Earth Laboratory
University of M'hamed Bougara
Boumerdès, Algeria

Yury G. Astrakhantsev
Head of the Laboratory of Borehole
Geophysics
Institute of Geophysics Ural Branch of
Russian Academy of Sciences (UB RAS)
Ekaterinburg, Russia

Eugenia Bazhenova
Junior Scientific Researcher
Laboratory of Borehole Geophysics
Institute of Geophysics Ural Branch of
Russian Academy of Sciences (UB RAS)
Ekaterinburg, Russia

Nadezhda A. Beloglazova
Senior Scientific Researcher
Laboratory of Borehole Geophysics
Institute of Geophysics Ural Branch of
Russian Academy of Sciences (UB RAS)
Ekaterinburg, Russia

Noureddine Djarfour
Faculty of Sciences and Technologies
University of Ahmed Draia
Adrar, Algeria

Vladimir S. Druzhinin
Lead Scientist
Laboratory of Seismic Research
Institute of Geophysics Ural Branch of
Russian Academy of Sciences (UB RAS)
Ekaterinburg, Russia

Mohammed Farfour
Assistant Professor
Earth Science Department
Sultan Qaboos University
Muscat, Oman

Jalal Ferahtia
Physics of Earth Laboratory
University of M'hamed Bougara
Boumerdès, Algeria

Said Gaci
Sonatrach–IAP
Boumerdès, Algeria

Olga Hachay
Professor, Lead Scientist
Laboratory of Borehole Geophysics
Institute of Geophysics Ural Branch
of Russian Academy of Sciences
(UB RAS)
Ekaterinburg, Russia

Galina V. Igolkina
Head of Laboratory of Exploration
Geophysics
Institute of Geophysics Ural Branch
of Russian Academy of Sciences
(UB RAS)
Ekaterinburg, Russia

Andrey Khachay
Associated Professor
Institute of Mathematics and Computer
Sciences
Ural Federal University
Ekaterinburg, Russia

Oleg Khachay
Associated Professor
Institute of Mathematics and Computer
Sciences
Ural Federal University
Ekaterinburg, Russia

Ignat Korchagin
Leading Researcher
Institute of Geophysics
Ukraine National Academy of Science
Kiev, Ukraine

Valery Korchin
Professor
Institute of Geophysics of Ukraine
National Academy of Science
Kiev, Ukraine

Sergey Levashov
Leading Researcher
Institute of Applied Problems of
Ecology,
Geophysics and Geochemistry
Kiev, Ukraine

Nikolay Nachapkin
Head of the Laboratory of Regional
Research
Institute of Geophysics Ural Branch
of Russian Academy of Sciences
(UB RAS)
Ekaterinburg, Russia

Orietta Nicolis
Institute of Statistics
Faculty of Science
University of Valparaíso
Valparaíso, Chile

Vjacheslav Osipov
Senior Scientific Researcher
Laboratory of Regional Research
Institute of Geophysics Ural Branch of
Russian Academy of Sciences (UB RAS)
Ekaterinburg, Russia

Daniyar Tazhibaev
Institute of Geomechanics and Mineral
Development of National Academy of
Sciences of the Kirghiz Republic
Bishkek, Kyrgyzstan

Kushbakali Tazhibaev
Professor
Institute of Geomechanics and Mineral
Development of National Academy of
Sciences of the Kirghiz Republic
Bishkek, Kyrgyzstan

Steven A. Tedesco
President, Consultant
Atoka, Inc.
Englewood, Colorado, USA

Nikolay Yakymchuk
Corresponding Member of the
National Academy of Sciences of
Ukraine, Professor
Management and Marketing Center of
Institute of Geological Science
NAS Ukraine
Kiev, Ukraine

Wang Jung Yoon
Geophysical Prospecting Lab
Energy & Resources Engineering
Department
Chonnam National University
Gwangju, South Korea

PREFACE

The goal of oil and gas exploration is to discover hydrocarbon accumulations that can be exploited in economic conditions. The geoscientists bring together information from various sources in order to evaluate the different petroleum system elements of a sedimentary basin. The book attempts to present different methods of oil and gas exploration, illustrated with worldwide case studies.

The book covers 16 chapters. Chapter 1 suggests a new technology of seismic geomapping based on new model conceptions of the upper part of lithosphere, and new depth criteria of forecasting deposits in new regions have been developed. Case studies from Russia illustrate the potential of the proposed approach.

Chapter 2 investigates a new statistical method based on fractal and multi-fractal analysis of Landsat 8 images for appraising the presence of mineral deposits and shale gas reservoirs. The potential of this method is shown through applications on different areas of study: northern Chile, the United States, and Argentina.

Chapter 3 aims at suggesting a denoising method using empirical mode decomposition. Applications on Algerian seismic traces demonstrate that the proposed method can serve as a good tool for denoising signals.

Chapter 4 is devoted to developing a lithological segmentation technique from well logs using the Hilbert-Huang transform, based on estimating a local scaling coefficient. This parameter measures the degree of heterogeneity of the layers crossed by the borehole. The proposed technique has been tested on synthetic well logs data and then applied on seismic velocity logs recorded at the KTB main borehole (Germany).

Chapter 5 introduces two free software packages used for VSP (vertical seismic profiling) data processing. The data processed using the suggested software are compared with those obtained from commercial software. The comparison demonstrates that free software packages can be utilized to process VSP data and produce results with quality that is comparable to that produced using commercial software.

Chapter 6 reviews the theory and the application of the time-frequency analysis or spectral decomposition on seismic data. The results obtained from southern Texas (USA) reveal features of the reservoir that are hidden in the seismic broadband.

Chapter 7 discusses the electromagnetic induction frequency geometrical method with controlled sources, a method that shows sufficient resolution and is constructed based on more complicated geological models. It is proved that this technique can successfully be used by oil production in mines.

Chapter 8 studies the reflection of processes of nonequilibrium two-phase filtration in oil-saturated hierarchic medium by active wave geophysical monitoring data. Using developed algorithms, it is possible to define the physical and structural features of a hierarchic oil layer structure and to estimate the water saturation from crack inclusions.

Chapter 9 deals with the definition of the surface of the fluid-saturated porous inclusion in the hierarchic-layered-block medium according to electromagnetic monitoring data. A three-stage approach, which is widely used for 3-D interpretation of mapping in the frame of the frequency-distance active electromagnetic method, is suggested for interpreting electromagnetic data.

Chapter 10 suggests a new technique based on three-component measurements of geoacoustic signals on oil and gas deposits to control the hydrocarbon deposit exploitation. The performance of the method is demonstrated on a case study taken from Russia.

Chapter 11 presents an application of borehole magnetometry for exploring oil and gas deposits in western Siberia. The use of such a method helps to understand the geological model and to perform an accurate deep structural forecast.

Chapter 12 introduces an original model to predict S-wave velocity integrating Hölderian regularity, empirical mode decomposition, and a multiple-layer perceptron artificial neural network (MLP ANN), from P-wave velocity logs. The obtained results demonstrate the effectiveness of the suggested model.

Chapter 13 examines a geophysical method, based on the law of variation of transverse waves velocity, for defining and checking the variation of stress in rocks at mining minerals. Examples of operating and residual stress definition of rocks are presented.

Chapter 14 discusses the possibility of using the mobile and direct-prospecting geophysical technologies to assess the prospects of oil-gas content in deep horizons. This technique has been implemented to evaluate the prospects of a number of oil-bearing areas and structures in the Dnieper-Donets basin and the Caspian basin.

Chapter 15 attempts to identify anomalies of low density in the crystalline crust of thermobaric origin based on laboratory study of the relationship between density and the longitudinal velocity of mineral material at high pressures and temperatures. It is shown that zones anomalies behavior of porosity and fractures parameters exist at different depths in the crystalline crust, and can play the role of canals for the migration and localization of hydrocarbons of deep origin.

Chapter 16 illustrates the use of aeromagnetic and micromagnetic surveys to identify potential areas of hydrocarbons in the midcontinental United States. Although the aeromagnetic survey identified basement features and might indicate fault systems that strongly coincide with existing petroleum fields, its application presents some shortcomings and its results are less definitive. The detected anomalous areas need to be evaluated by using other surface geochemical methods. Additional methods such as subsurface geologic mapping, drilling, or seismic are also helpful to define whether these anomalies are real or false.

Overall, the book covers different aspects of oil and gas exploration and will be of a high interest for practitioners and academicians. We hope you will appreciate this book.

Said Gaci
Olga Hachay

ACKNOWLEDGMENTS

We would like to express our appreciation to all the authors who submitted valuable chapters, particularly Dr. Mohamed Farfour and Dr. Orietta Nicolis for their precious advice to improve the quality of the book.

Special gratitude goes to Ms. Madiha Gaci for her sustained help in proofreading the book.

We are grateful to Mrs. Ouiza Gaci for her patience, understanding, and encouragement.

Our sincere thanks and gratitude are also expressed to the Algerian Petroleum Institute for the continued support while working on this book.

We would like to thank the editorial staff of Wiley for their concise and thorough work. We deeply appreciate the assistance that Ms. Rituparna Bose has given us throughout the process of bringing this book together.

1. EXPERIENCE OF REGIONAL PREDICTION OF HYDROCARBON DEPOSITS PROSPECTING IN THE VICINITY OF URAL OIL AND GAS PROVINCES

Vladimir S. Druzhinin, Vjacheslav Osipov, and Nikolay Nachapkin

Abstract

This chapter is devoted to investigating oil and gas content in the Ural region according to the position of the Earth's crust deep structure. A significant volume of deep seismic soundings was provided in the Ural region by the Bagenov geophysical expedition, by the Institute of Geophysics (Ural Branch of Russian Academy of sciences), and by the Center of Regional Geophysical Research (GEON) during some decades. We have elaborated a new technology of seismic geomapping, based on new model conceptions about the structure of the upper part of lithosphere. We supplement the base of seismic research by the geological data (information about the oil and gas content of the region), the data of super deep boreholes, and area gravimetric data. That allowed us to construct a geological-geophysical fault-block volume model of the lithosphere (up to 80 km), to provide the tectonic mapping according to these cuts: the surface of ancient crystalloid crust and the main seismic geological surface M. On the basis of the obtained information compared with the data of hydrocarbon deposits location, new depth criteria of forecasting deposits in new regions and objects have been developed. We can assign some examples of positive forecasting. They are the Kueda region (south of Perm region) and the region of the town Khanti-Mansyisk.

Institute of Geophysics Ural Branch of Russian Academy of Sciences (UB RAS), Ekaterinburg, Russia

Oil and Gas Exploration: Methods and Application, Monograph Number 72,
First Edition. Edited by Said Gaci and Olga Hachay.
© 2017 American Geophysical Union. Published 2017 by John Wiley & Sons, Inc.

1.1. Introduction

In the last decade, interest in the information of Earth's crust and upper mantle structure has steadily grown. That knowledge has become required for developing deep submerged sediments, which are the main source for refueling the world base of oil and gas of the 21st century, and for prospecting objects in new areas. Success of prospecting new fields of hydrocarbon (HC) deposits inside the territory of oil and gas areas of the Ural region is too low. One of the main reasons for this may be that the peculiarities of the deep structure chosen for detailed exploration are not usually taken into account. That situation is caused by the lack of sufficiently reliable data on the structure of the Earth's crust, as well as absence of criteria for the prediction of prospective areas. A simplified system of observation in deep seismic sounding oriented on the subhorizontal environment is in conflict with the complex block model of Earth's crust, established by using the dense continuous profiling systems and the results of deep drilling. First of all, we need to have information about the tectonic model of the upper lithosphere and also about the connection between the structural and tectonic elements of the sedimentary cover, including tectonics of crystalline crust and of deep fluid dynamic zones. That will determine the types of the Earth's crust as a generation source of hydrocarbons that contributes to their vertical migration into upper layers of the geological environment. The development of depth criteria that determine the formation of large hydrocarbon deposits was introduced in the 1960s. In the beginning it was realized in profile models, and then in area systems for constructing a geological-geophysical volume model of the Ural's upper lithosphere and scheme of tectonic districts of the crystal crust. Conclusions about the prospects (or nonprospects) of areas were confirmed by subsequent detailed prospecting operations. The need for such an approach at the stage of regional and regional-zonal forecast in this chapter is considered an example of research on oil and gas provinces of the Ural region.

1.1.1. Some Information About Oil-Gas-Reserves of Ural's Region and Their Geological Environment

Overviews have been published in the works by *Bochkarev and Brechuntsov* [2008], *Gogonenko et al.* [2007], *Dontsov and Lukin* [2006], *Megerja* [2009], and *Timonin* [1998]. On the Ural region territory (48°–70° east latitude, 56°–68° north longitude), there are three large oil gas provinces (OGP). The most important one, by volume of developing hydrocarbon minerals, is western Siberia, followed by the Timan-Pechora and Volga-Ural OGPs. The main productive horizons in the western Siberian OGP are located in the Mesozoic interval at the depth 1.2–3.5 km, which is well studied by geologic and geophysical methods, and therefore discovery in the limit of known oil gas regions of large and super large HC deposits is unlikely. In the opinion of leading scientist-oilers, a high capacity can define some negative structures of the near Ural part of the western Siberian plain—for instance, the Ljapinsky depression. As a whole, the resupply of the resource base of HC in the western Siberian OGP consists of the deeper horizons, first from the sub platform sediments of the Triassic-Devonian age and in the disintegrated part of the basement (before Jurassic base). The research in that direction began in the mid-1980s, but unfortunately there were not any significant results, only some objects corresponding to small oil deposits. Timan-Pechora OGP differs by a widely known age interval of oil gas content of sedimentary rocks (from Silurian to Triassic, inclusive) and by a significant role of faults in the HC deposits location. That province is very well studied, except the eastern and southern regions. Namely, these territories are regarded as prospective for HC deposits prospecting. Regional-zone searches for HC, which have been conducted in these regions, haven't given any results yet. An actual problem of searching for HC in deeper areas of known productive horizons is encountered in the whole Volga-Ural OGP, including for Perm and Bashkiria OGP, and on the eastern part of that region into the inner part of near Ural depression and in the western side of the Ural fold system.

The main cause of noneffectiveness of prospecting in the shown direction according to our opinion is that orientation in these works on the sedimentary-migration theory of HC generation and deposits formation, using that as a base of the existence of oil deposits. An example is the deposit located in the pre-Jurassic basement of western Siberia. It occurs due to horizontal migration of HC from the productive horizons of sedimentary basins. The next causes are related to ignoring the specificity of the deep structure and mainly vertical migration, which are the basic elements of developing a new paradigm of oil geology [*Dmitrievskiy*, 2009; *Timurziev*, 2010; *Dmitrievskiy*, 2012; *Bembel and Megerya*, 2006]. At last we see the imperfection of searching works technology, which does not take into account the complicated medium structure and nontraditional character of HC accumulation location.

1.2. Method of Information Analysis for the Regional Prediction

During 2006–2011 in the Institute of Geophysics, the research work of developing a method and creating a volume geologic-geophysical model of the upper part of the Ural region lithosphere had been provided for the territory with coordinates 54°–68° n. l. and 50°–71° e. l. The main initial data had been of the same type of seismic geological sections of deep seismic sounding (DSS) profiles (geotraverses), developed by using the provided technology of seismic mapping of Earth's crust [*Druzhinin et al.*, 2013b, 2014a] and gravitational field in Bouguer reduction. A scheme of profile locations combined with the field Δg is shown in Figure 1.1.

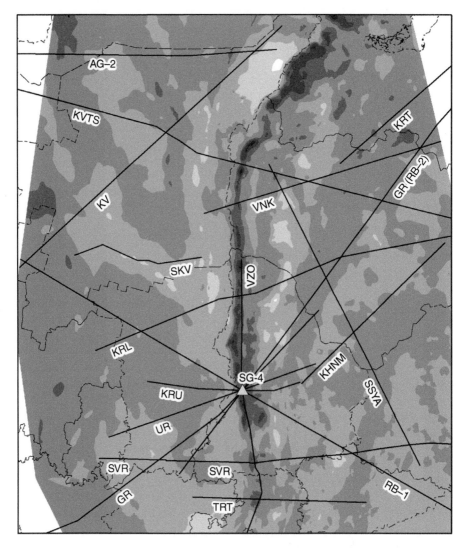

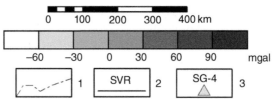

Figure 1.1 Location map of deep seismic sounding (DSS) and method of exchanged waves (MEW) profiles, combined with a map of anomaly gravitational field Δ*g*.
Legend: 1 = administrative borders of RF regions; 2 = Rubin-1; RB-2 = Rubin-2; UR- Ural MEW; KRT = Kraton. Profiles of Bagenov geophysical expedition with participation of the Institute of Geophysics UB RAS: SVR = Sverdlovsk; GR = Granit; TRT = Taratash; KRU = Krasnouralsk; HNM = Hanti-Mansian; SSJA = Northern Sosva-Jalutorovsk; KRL = Krasnoleninsk; VNK = Vernenildino-Kazim; VZO = Vizaj-Orsk. Profile of Spetsgeofizika: KV = Kupjansk-Vorkuta. Profile of the Institute of Geology Komi Center UB RAS: SKV = Syktyvkar; 3 = Ural super deep borehole.

In the first step of that method, the seismic section was used for the sublatitude Sverdlovsk intersection of the Ural 1000-km length, which was developed according to materials of continuous profiling and systems of hodographs of refracted (weak refracted) waves with a length up to 300 km, wide refractions with an interval of registration from 0–20 km to 150–200 km. On the Ural region, we acquired 15 seismic sections of the Earth's crust with total amount more than 10,000 km. In Figure 1.2a, b we can see as examples sections for the region of the super deep borehole (sublatitude Krasnouralsk and meridional Vizhai-Orsk).

For more details, the technology of developing the volume geologic-geophysical models is written in the following papers and monographs: *Druzhinin* [1978, 1983, 2009], *Druzhinin et al.* [1976, 2004, 2009, 2010, 2012, 2013a, 2013b, 2014a, 2014b], and *Martyshko et al.* [2011]. An example is shown in Figure 1.3.

1.3. Depth Criteria by Estimation of Oil Gas Potential of Geological Medium

During the process of profile investigation using DSS, two directions of the regional prediction were defined. The first direction was suggested by *Bulin et al.* [2000]. It consists of possible control of the known HC deposits region by values of velocity parameters of longitudinal and transversal waves for bottom domains of the Earth's crust, which were defined by the results of many wave point soundings on the geotraverses of the Center of Regional Geophysical Research (GEON). The main meaning of that analysis was given to the contrasts of the velocities. The difference of the velocities of the longitudinal waves with meanings 0.2–0.4 km/s and the values of the ratio Vp/Vs 0.05–0.2 were taken into account. By the use of more simple systems of observations on large profiles these values of the parameter differences are often on the mistake-defining level. Furthermore, the structure-tectonic factor was not considered, because the structural differences were very small on the deposit places and out of it. Therefore, we can recognize that by researching the geotraverses for the examples of large HC deposits, the possibility of using much wave seismicity for regional prospecting (allocation prospective areas with linear dimensions up to 100 km) has been shown. But by using that data, it may be possible to consider places for drilling search-parametric boreholes, as shown by *Bulin et al.* [2000]. The second direction is oriented on the structural-tectonic factor. By the way, the specific peculiarities of the Earth's crust structure and upper mantle in the region of known HC deposits are transferred onto other objects of the OGP. That direction is based on the results of a profile research with sufficient dense systems of observation, provided by the Bagenov geophysical expedition together with the Institute of Geophysics Ural Branch of the Russian Academy of Science (UB RAS) [*Druzhinin*, 1983]. The morphological features are supplemented by tectonic features (deep faults, density of dislocations with a break of continuity and their

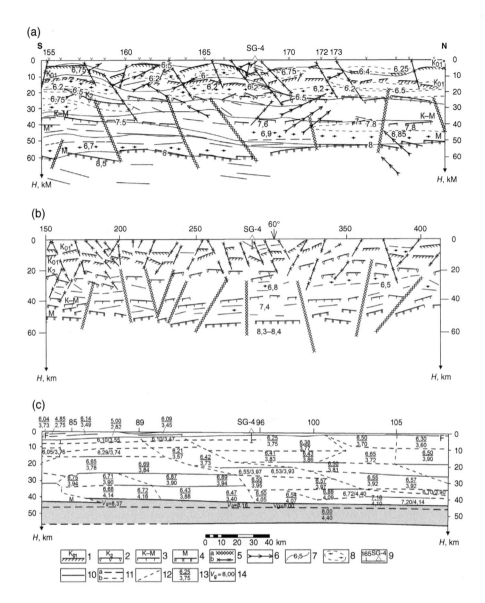

Figure 1.2 Legend: 1–4 = section surfaces of seism geological floors, which were constructed according to a set of seismic data, including the schematic velocity section: 1-II (SGE), or of the ancient crystalline basement, 2-III (SGE), or protocore; 3 = transition mega complex at the bottom of the crust; 4 = the main seism geological section M; 5 = assumed zones of deep faults (a) and deep reflected elements, which take a secant position to the main structural plan; 6 = the reflected structural elements; 7 = isolines of the volume section, developed mainly according to reflected subcritical waves, km/s; 8 = assumed zones of velocity inversion; 9 = line of the profile together with pickets, their numbers, and projection of the Ural super deep borehole SG-4; 10 = seismic boundaries in the crust and upper mantle, developed according two types of waves and more; 11 = for a mono type wave field; 12 = block boundaries with different velocity parameters; 13 = value of the longitudinal (in the numerator) and transversal (in the denominator) values, km/s.; 14 = edge velocity according to the wave P^M.

(a)

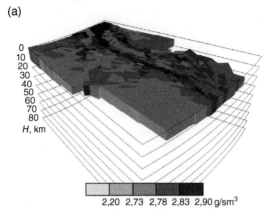

2,20 2,73 2,78 2,83 2,90 g/sm^3

(b)

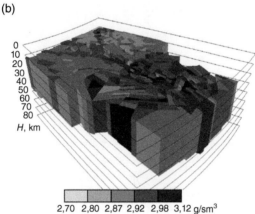

2,70 2,80 2,87 2,92 2,98 3,12 g/sm^3

(c)

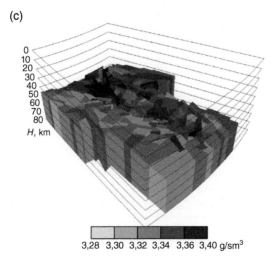

3,28 3,30 3,32 3,34 3,36 3,40 g/sm^3

Figure 1.3 Density 3-D model of the northern and central parts of the Ural region: a-layer $K_0 - K_{01}$; b-layer $K_{01} - M$; c-layer M-80 km [*Druzhinin et al.*, 2014a, pp. 66–69].

character on different levels of the Earth's crust). Besides, the contrast of the values of physical parameters of the Earth's crust blocks and the type of the deep structure is taken into account here. We can see below the concrete examples of such a prediction. Using the volume-density model of the upper part of the lithosphere (UPL) (Figure 1.3), the tectonic zoning of the researched territory of the Ural region was done by dividing into three large partitions: upper part (first seismological level: K_0–K_{01}), crystalline crust (from the surface of the ancient basement up to the surface of the main seismological section M: K_{01}–M), and an upper part of the upper mantle (from M up to regional level of the isostatic alignment –80 km: M–80 km). The scheme of crystalline crust zoning, which is combined with known HC deposits, is shown as an example in Figure 1.4. According to the analysis of this data, the main part of the deposits is linked with the deep faults of the UPL and with the places of their crossings. Among these deposits, we can divide "productive" fault zones, which are linked with the chains of HC deposits. We can attribute them to the fault structures of the Pechora rift system, the Kaltasin aulacogen of the Volga-Ural block, and the edge stitch along the eastern boundary itself, the Uralides.

A great role in the location of known deposits possibly belongs to deep shear dislocations, which can be seen by displacements of the block contours that confine faults and link them to HC deposits in oil regions. They are the Bardim area, the Pechora rift structure (Figure 1.4), and the Hanti-Mansyrian region (Figure 1.4). All these structures are characterized by large HC deposits. This conclusion was made by both Gogonenko and Timursiev for the pre-Jurassic bottom of the western Siberian OGP [*Gogonenko et al.*, 2007; *Timursiev*, 2010]. It is not dismissed that the mainly elliptical shaped concentration of the basic oil region deposits and their confinement to the places of deep fault intersections, show on their possible confinement with the deep fluid dynamic zones of the UPL. The type of deep structure, the dominant content of the consolidated crust, and the character of structures' evolution are significant to the question of regional prediction. For example, the prospecting of HC deposits is not probable

Figure 1.4 The location map of oil and gas prospect estimation of Ural region geological medium according to specificity of the upper part lithosphere structure.
Legend: geoblocks: VUGB = Volga-Ural, MBGB = Mezen-Belomorsk, USS = Ural faulted structure, ZSGB = western Siberian, KZGB = Kazakhstan. 1. Contours of assumed paleoactive structures in the upper part of lithosphere (1 = Kaltasin rift—aulacogen, 2 = Kirov-Kagim aulacogen, 3 = Central-Ural mega zone, 4 = section slip-eastern boundary USS, 5 = section slip between VUGB and MBGB, Central Pechora mega zone; 6 = Pechora rift zone). 2. Contours of geoblocks (a) and mega zones (b). 3. Fault tectonic: sublatitude dislocations (a) and deep faults (b). 4. Contour of the Ljapin mega trough. 5. Areas in the limits of known oil and gas regions, which are prospective for searching HC deposits mainly in more deep horizons. 6. Other prospective areas. 7. Contour of Kueda area, which is prospective for searching HC into the Kaltasin complex early Riphean age. 8. Deposits of hydrocarbons.

in the regions of the Ural central mega zone, which belongs to the paleorift structure with a crust of femic profile, though there are sediment basins (Chelyabinsk graben, Mostovskaja and Karpinskaja depressions, Ljapinsky mega trough [*Druzhinin*, 2009], and others).

To sum up, we should emphasize that to increase the level of the regional prediction of prospecting HC deposits, it is necessary to use information about the deep structure of the geological medium, combining the two considered directions, but the decisive method is the tectonic zoning of the UPL using the technology of constructing the geology-geophysical 3-D model. In this chapter we do not consider such defining factors for formation of HC deposits as the age of deposit formation, existence of the collectors and covers in the sediment layers, intermediate mega complex, and disintegrated part of the bottom of sediment basins. For that, we must provide a detailed stage of prediction on a minimum scale of 1:200,000 on the prospective areas, which are distinguished in the first stage of the research using analysis of the geologic-geophysical information. Only after that can we begin to provide a detailed prospecting on the promising sites in the area.

-->

Figure 1.5 Examples of estimation of oil and gas potential of geological medium with use of profile DSS data, which has been confirmed by results of prospecting and exploration: a, b: positive estimation; c, d: negative estimation.
Legend: a = Hanti-Mansian profile, Frolov depression (I), b = Sverdlovsk intersection, Kaltasin avlacogene (II),c = Sverdlovsk intersection, Vagaj-Ishim depression (III), d = Troitsk profile, South Ural, Bashkiria anticlinoria. Seismic geological sections of Earth's crust(1-6): 1 = upper part of the first seism geological level (SGL), presented by Paleozoic sediments (PZ) (VEP); mezokaynozoy of Western Siberia (MZ-KZ) (a); folded Paleozoic complexes (b); Riphean, mainly terrigenous Riphean sediments of Western Ural (c); 2 = bottom complex I (SGL), probably metamorphic rocks of the basic content of low Proterozoic; 3 = a second SGL, mainly granite-gneiss, gneissic rocks of the Archean basement; 4 = the third SGL corresponding substantially reworked during the evolution to rocks of species protocrust; 5 = the transition complex in the lower crust, which structure may contain ultramafic intrusions, gabbro, basic granulites, anorthosite; 6 = the main Seismogeological section M, corresponding to a relatively weakly altered rocks of the upper mantle; 7 = dolomite Lower Riphean complex Kaltasinsky aulacogene with Vp = 6,8 km/s; terrigenous shale complex of Lower Riphean Kaltasinsky aulacogene; 8 = reflective elements within the Earth's crust and upper mantle; 9 = values of the velocity of longitudinal waves in km/s; breaking tectonics: 10 = deep faults and fault zones (a), faults in the upper layers of the crust (b); 11 = an intermediate complex in the lower sedimentary layer of Western Siberia represented Triassic basalts and clastic-slate rocks of the Upper Paleozoic-Triassic; 12 = the top of the complex protocrust represented by gneisses, amphibolites, gneiss, amphibolite, gabbro; 13 = reflecting elements occupying intersecting position with respect to the basic structural plan; 14 = local areas of the medium corresponding to the dynamically active elements; 15 = the border zone between the Ural deflection and West Ural structure; 16 = predictive search and parametric wells: 1. Khanty-Mansiysk (forecast Druzhinin, 1983), 2. Kulguninskaya deep well (H = 5.2 km); 17 = promising area; 18 = settlements: 1. Kueda, 2. Berezovka.

1.4. Discussion of the Results

1.4.1. Estimation of Oil Gas Prospecting by Profile Works

As a confirmation of the role of deep factors, in Figure 1.5a, b we have shown the seismological sections for the places of the DSS profiles, on which subsequently, independent from given recommendations, geological and geophysical

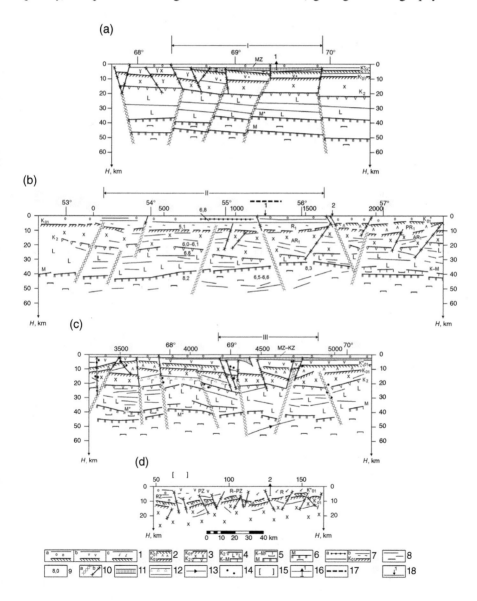

works have been carried out for searching HC deposits. Figure 1.5a shows the section of the eastern end of the Hanti-Mansian profile [*Druzhinin*, 1983] where the drilling of a parametric borehole was recommended for confirming the existence of the sediments of the intermediate complex on the median Hanti-Mansian elevation for estimation of its prospecting. The area is named Gorelaja area. Our prediction in 1985 was confirmed. Drilling revealed an anomaly of the Earth's crust on the western part of Sverdlovsk intersection (Figure 1.5a): a large elevation of the surface M, an increased thickness of the sediments due to lower Riphean complex, and diminished basicity of the crystalline crust, which changes to the east of the Chernushen fault by a block of increased basicity [*Druzhinin et al.*, 2010, 2013b].

Therefore, an assumption has been made about a large prospect of the region Kueda, to the west from the relatively small deposits Chernushki-Beresovka. At the present time that region is the largest oil extraction region (southwestern Perm region and the neighboring area of the republic Bashkortostan). Using the DSS data on the east of the Sverdlovsk intersection, the Vagaj-Ishim depression was discovered, probably filled by the intermediate complex of the Devonian-Triassic age (Figure 1.5b).

But the increased basicity of the crystalline crust (the velocity values and calculated density are increased) did not allow for distinguishing that area for oil prospecting. Another example is related to the western slope of the South Ural. A deep (>5.0 km) borehole was drilled within the Bashkiria elevation, which was folded by Riphean rocks, by opening the Paleozoic productive horizons of the eastern edge of the East European platform, which was expected in the shifted location under the Riphean complex R_1 with a thickness of 4.5 km. The thrusts were assumed to be very significant: up to 70 km. But nothing of that is seen in Figure 1.5d. The section of the DSS Troitsk profile, which corresponds to the upper part of Earth's crust, is broken by tectonic faults of opposite directions and complicated relief of the assumed surface of the ancient crystalline basement, located in the interval of depths 5–14 km. In that area the depth of the Kulgunin borehole is 7–8 km.

With the help of drilling data, the normal section of Riphean sediments is obtained; their summarized thickness was in good agreement with depth of the ancient crystalline basement. The deeper boreholes or their shifting nearer to the boundary zone of the place, which is shown in Figure 1.5b, unlikely clarify the situation with the sub thrust oil in the western Ural mega zone and at least in the Bashkiria anticlinoria in the southern Ural. We would like to emphasize: we could come earlier to that conclusion by using an objective analysis and taking into account the information about the deep structure of that region.

1.4.2. Estimation of Oil and Gas Prospecting of Geological Medium of Main Oil and Gas Provinces and Area of the Ural Region on the Basis of Obtained Information and Revealed Depth Criteria

1.4.2.1. Timan and Pechora Oil and Gas Province (northwest part, Figure 1.4), Limited on the East by 62° and on the South by 62° of Northern Latitude

For this area, the perspectives of oil and gas prospecting had been extended due to the western mega block; a chain of prospective places is located along the contact (deep fault) Izma-Pechora block with the Omra-Luzkaja rift zone. The promising areas are located on the intersection of longitudinal faults with the sublatitudinal dislocations and diagonal faults. In the limits of the Pechora mega block, a number of areas have been mainly outlined for prospecting deposits in deep horizons. The perspectives of the eastern mega block (Chernishevskiy and Rogovo-Kosvinskiy) are not finally distinguished. We had to take into account the analysis of the oil- and gas-promising Pechora mega block features, that the activity of dynamic processes, which are linked with the origin of the rift system, were continued in Trias (basalts effusion), the Neokomian reversal of the structural plan from a north-northwestern to north-northeastern direction, which led to shear dislocations. Probably the activity of the Pechora mega structure is saved on the current stage, which is a positive moment by the estimation of perceptiveness of that structure.

1.4.2.2. The Eastern Part of the Volga-Ural Oil and Gas Province of the East European

Platform, (the south-eastern part, limited on the east by meridian value of 58°, Figure 1.4)

The most promising region, according to information of the deep structure, is the Kueda region—the southwestern part of the Perm territory (Figure 1.6a). It is located in the limits of an assumed deep fluid dynamical zone, which was formed at the intersection of deep diagonal faults and which is probably accompanied by shear dislocations. In geological plane to the Paleozoic sedimentary cover to that region, the articulation of the Kaltasin aulacogen and the Kama-Carnelian trough is located. In their limits, the lower Riphean dolomites can contain oil, which are located on relatively not large depths of 2.5–2.7 km. Due to a significant fracturing of certain areas, which tend to deep faults, the existence of a fractured reservoir is not excluded. The relatively smaller depth of dolomite strata on the southwestern side of the Perm territory and on the northern side of the Bashkiria Republic can significantly facilitate the prospecting of such type of deposits. The dolomite sediments can also be promising in that region. The promising areas are

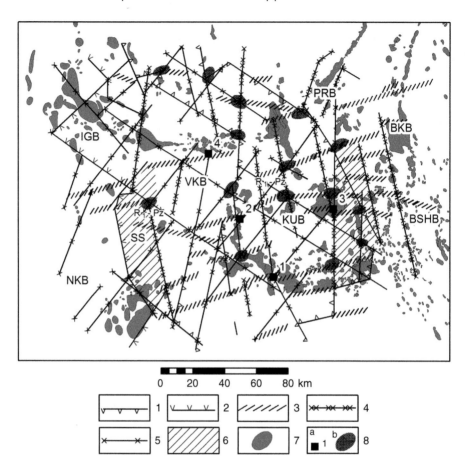

Figure 1.6 Scheme of potential oil and gas surveys of low Riphean complexes, located in the southwestern Perm territory (Bardim area).
Legend: Blocks of the Earth's crust: IGB = Izhevsk, NKB = Nizhniy-Kamskiy, VKB = Votkinskiy, KUB = Kuedinskiy, PRB = Permskiy, BKB = Bimsko-Kungurskiy, BSHB = Bashkiria, SS = Sarapul-structure. Boundaries of potential prospecting areas: 1. Kueda; 2. Sarapul-structure and Izhevsk-block, disjunctive tectonics; 3. Sublatitude boundary zones; 4. Subvertical faults; 5. Elements of shear tectonics; 6. Border areas; 7. Known oil deposits [Neganov, 2010]; 8. Prospecting areas (1–4) for arranging detailed research of the first stage (a) and next stage (b).

outlined outside the borders of the contour shown in Figures 1.4 and 1.6, and they can be located in the limits of the considered net of deep orthogonal faults.

1.4.2.3. The Northern Part of the Perm Territory

The prospect of the northern part of the Perm territory is not significant, because according to the scheme of tectonic zoning (Figure 1.4), it is located in the lower Proterozoic folded complexes of mafic profile, which are analogous to the karelides of the Baltic shield. The interest of HC processing was shifted to certain areas of the Kazim aulacogen and sublatitude dislocation in the region of the 61st parallel. The region of intersection has to be classified into three large structures of the Ural region: Volga-Ural, Timan-Pechora, and the Ural folded system. Besides, the Paleozoic complexes here can be as pre-Paleozoic oil saturated sediments. The prospects of the top-folded zone of the western Urals slope are limited, excepting two or three areas, which coincide with "pockets" that are formed by fault-slip dislocations. That region of the Sabik area is defined as promising according to results of generalization of deep seismic soundings data and constructing of the geological-geophysical volume model for the Shalja-Mihailov area [*Druzhinin et al.*, 2004].

1.4.2.4. Results of Parametric Drilling on the East Ural

According to parametric drilling in 2011 in the east Ural, a significant amount of gas was determined in the productive eastern part of the East European platform carbon horizons, which has here an increasing thickness, compared with other regions of the eastern part of the Pre-Ural trough. The same situation can be also similar to the north areas. The suggestion about the small promise of the top-folded zone is contrary to known results about a high promise of that territory. To support the thesis with another example from the south Ural, let us indicate two characteristic features of the western slope of the Ural: first, the large plates, which were created by horizontal movements, are absent or poorly developed—they are mainly shifted structures with small horizontal amplitudes; second, the geological medium of the western Ural mega zone as a compound part of the Ural folded system has a complicated structure and has, especially the eastern part, a mainly metallogenic potential.

1.4.2.5. Oil and Gas Province of the Western (Pre-Ural) Area (Western Siberia) (Figure 1.4)

The southern segment of the western Pre-Ural is the most promising; its geological medium contains significant oil and gas potential, indicated by the existence of many super large, large, and middle HC deposits. This promising band, which has been defined and which contains no fewer than 10 areas, coincides with the boundary seam and Hanti-Mansian and Nadym ancient massif of the western Siberian geo block. The method of prediction was the same, as for the Timan-Pechora oil and gas province. By defining the prospect areas, we took into account the probable confidents of yet known deposits to probable deep shear dislocations, especially significant in the limits of the Hanti-Mansian mega block. Some areas are defined behind the limits of the oil regions, but the main part is revealed in these regions and oriented mainly for searching HC deposits in the intermediate complex and in the disintegrated part of the bottom of the sedimentary basin. To the south of the 60th parallel on the continuation of the band, some prospective areas have been revealed. The prospects of the eastern Ural mega zone and the eastern part of the central Ural mega zone (SUMZ) are limited in spite of the existence in the northern part of sufficient thicknesses of sediments of the Mesozoic-Cenozoic age (more than 1–1.5 km). The high prospect of the Ljapinsky trough, which is assumed without considering the specificity of Earth's crust structure, was not confirmed by further geological-geophysical works. It had not been taken into account the conclusions about the small potential of that structure, which had been made by providing DSS, profile Verchnenildino-Kizim in 1986 and then by constructing the volume geological-geophysical model UPL for the northwestern part of the eastern Siberian plane [*Druzhinin*, 2009]. As reciprocally prospect areas, some areas have been defined along the east contact of the SUMZ in the limits of the neighbor strip of elevations of the eastern Ural mega zone (Figure 1.6). Here we do not give information about the areas' surveys, which are confined to a granite massif of upper Perm age, which is given in *Druzhinin et al.* [2014].

1.5. Conclusions

The new technology of regional prediction of hydrocarbon fields and the solution of tectonic problems of regional oil and gas geology on the basis of lithosphere upper structure information is offered. We forecasted more than 10 prospective sites for statement of detailed exploration. This technology concerns immersed deposits in known oil areas and focuses on new areas. Positive results of good orientation exploration will lead to a significant increase in the resource base of hydrocarbons in the oil and gas provinces of the Ural region. Research on the creation of geologic-geophysical models developed at the Institute of Geophysics UB RAS needs to be continued. Research should focus on all territory of the western Siberian platform and Timan-Pechora OGP, including the Pechora Sea; other areas include earlier Phanerozoic and Paleozoic complexes, and perhaps productive deposits of the western slope of the Urals and regions of granite massifs. It is important to increase reliability and detailed information on deep structures through interpreting results on regional profiles of deep OGP, expanding geophysical methods, and obtaining data on superdeep drilling. For promising areas, it is necessary to perform additional analysis of the obtained information together with geological and geophysical data on the near-surface structures with participation of experts for each province.

References

Bembel, S., V. Megerja, et al. (2006), Prospecting and exploration of hydrocarbons on the basis of the concept geosoliton degassing of the Earth. *Geology of Oil and Gas*, 2, 2–29 (in Russian).

Bulin, N., et al. (2000), Prediction of oil petroleum potential according to deep seismic profiles. *Regional Geology and Metallogeny*, 10, 195–204 (in Russian).

Bochkarev, V., and A. Brechuntsov (2008), Generalized tectonic models of the west-Siberian geosynclines. *Gornie Vedomosti*, 3, 6–23 (in Russian).

Dmitrievskiy, A. (2009), *Selected Works, Vol. 2. Fundamental problems of Earth's sciences*. Nauka, Moscow (in Russian).

Dmitrievsky, A., and B. Valyaev (eds.) (2012), *Degassing of the Earth and the Genesis of Oil and Gas Fields*. Geos, Moscow (in Russian).

Dontsov, V., and A. Lukin (2006), About endogen factors of oil deposits forming into the crystalline basement of Kiulongian depression into the shelf of South Vietnam. *DAN*, 11,10–18 (in Russian).

Druzhinin, V. (1978), Characteristics of the Ural deep faults according to seismic data. *Soviet Geology*, 4, 146–154 (in Russian).

Druzhinin, V. (1983), The features of the deep structure of West-Siberian plate along the Hanti-Mansian DSS profile. *Geology and Geophysics*, 4, 39–45, (in Russian).

Druzhinin, V. (2009), Information of Earth's crust structure is necessary attribute of regional oil geology (according to the example of Ural region). *Russian Geology*, 6, 65–70 (in Russian).

Druzhinin, V., V. Kolmogorova, et al. (2009), The map of pre-Yurian matter complexes in the northwestern part of western Siberian plain on the base of the volume model of the Earth's crust, *Russian Geology*, 1, 104–112, (in Russian).

Druzhinin, V., P. Martyshko, et al. (2012), Solution of the problems of regional oil geology of the middle segment of the Ural region on the base of volume geological-geophysical model of the upper part of lithosphere. *Geology, Geophysics and Exploration of Oil and Gas Deposits*, 1, 32–41 (in Russian).

Druzhinin, V., P. Martyshko, et al. (2013a), The scheme of tectonic zoning of the Ural region on the base of information about the structure of the upper part of lithosphere. *Russian Geology*, 1, 43–58 (in Russian).

Druzhinin, V., P. Martyshko, et al. (2013b), Regional structural features of Lower Riphean sections in connection with the problem of petroleum resource potential of deeply buried complexes in the southwestern Perm region, *Doklady Academia Nauk*, 452 (2), 181–184.

Druzhinin, V., P. Martyshko, et al. (2014a), *Structure of the Upper Part of the Lithosphere and the Oil Petroleum Potential of the Ural Region*. IGF UB RAS, Ekaterinburg, (in Russian).

Druzhinin, V., N. Nachapkin, and V. Osipov (2010), The role of information about the structure of the Earth's crust for tectonic zoning and estimation of oil prospecting of new regions and new objects. *Geology, geophysics and exploration of oil and gas deposits*, 11, 10–18 (in Russian).

Druzhinin, V., V. Osipov, and A. Pervushin (2004), About HC searching in the southwestern part of the Sverdlovsk region. *Prospecting and Protection of Natural Resources*, 2, 29–30 (in Russian).

Druzhinin, V., V. Rakitov, et al. (2014b), Crustal structure in the Polar sector of the Urals folded system (from DSS data). *Russian Geology and Geophysics*, 55, 390–396.

Druzhinin, V., I. Sobolev, and V. Ribalka (1976), *The Link of Tectonics and Magmatism With the Deep Structure of the Middle Ural According to DSS data*. Nedra, Moscow (in Russian).

Gogonenko, G., A. Kashin, and A. Timursiev (2007), Horizontal basement shifting of western Siberia, *Geology of Oil and Gas*, 3, 3–11 (in Russian).

Martyshko, P., V. Druzhinin, et al. (2011), Method and results of volume geological-geophysical model development of the lithosphere upper part of the northern and middle segments of the Ural region. In *Dynamics of the Earth's Physical Fields*. IPHE RAS, Moscow, pp. 9–30 (in Russian).

Megerja, V. (2009), *Search and Exploration of HC Deposits, Which Are Controlled by Geosoliton Degassing of the Earth*. Locus Standy, Moscow (in Russian).

Neganov, V. (2010), *Seismological Interpretation of Geophysical Materials on the Middle Urals Region and Prospects of Further Petroleum Investigations*. Perm University, Perm (in Russian).

Timonin, N. (1998), Pechora plate: History of geological evolution in the Phanerozoic. Institute of Geology Komi Scientific Center UB RAS, Ekaterinburg (in Russian).

Timursiev, A. (2010). The modern state of practice and methodology of searching oil: From misconception stagnation to new ideology of progress. *Geology, Geophysics and Exploration of Oil and Gas Deposits*, 11, 20–32 (in Russian).

2. WAVELET-BASED MULTIFRACTAL ANALYSIS OF LANDSAT 8 IMAGES: APPLICATIONS TO MINERAL DEPOSITS AND SHALE GAS RESERVOIRS

Orietta Nicolis

Abstract

The aim of this work is to explore a new statistical tool based on fractal and multifractal analysis of Landsat 8 images for assessing the presence of mineral deposits and shale gas reservoirs. The method uses the 2-D discrete wavelet transform for evaluating the fractal and multifractal spectra, and produces some descriptors that could be useful for discriminating mineralized and shale gas areas. For showing the performance of the method we consider three different areas of study: an area located in the North of Chile, a wide shale gas reservoir in the United States, and a shale gas area in the Vaca Muerta region of Argentina.

2.1. Introduction

The use of satellite images in hydrocarbons exploration has been recently employed for the identification of structures with potential for traps formation and the detection of potential areas for hydrocarbon microseepages on the surface as it can be associated to some geotectonic features such as fractures and faults [*Barberes*, 2015]. Many studies indicate that the distribution of petroleum resources exhibits a self-similar characteristic, in both pool size distribution and spatial geometry [*Barton et al.*, 1991; *Chen et al.*, 2001]. This feature has motivated the

Institute of Statistics, Faculty of Science, University of Valparaíso, Valparaíso, Chile

Oil and Gas Exploration: Methods and Application, Monograph Number 72,
First Edition. Edited by Said Gaci and Olga Hachay.
© 2017 American Geophysical Union. Published 2017 by John Wiley & Sons, Inc.

analysis of the fractal approach for describing the spatial distribution of the hydrocarbon resource. The fractal model has been used by many authors for assessing the aggregate properties of petroleum accumulations and for estimating the pool/field size distribution (see e.g., *Mandelbrot* [1962], *Drew et al.* [1988], *Houghton* [1988], *Crovelli and Barton* [1995], and *La Pointe* [1995], among others).

In this work we propose a new approach based on the wavelet and multifractal spectrum for assessing the presence of mineral deposits and shale gas reservoirs. The method is based on fractal and multifractal analyisis of a series of Landsat 8 images taken in different part of the world where mineral deposits (especially copper and gold) and shale gas reservoirs were met. The multifractal analysis allows us to estimate some descriptors that we think might be useful for characterizing highly mineralized and shale gas areas.

Multifractal analysis is concerned with describing the local singular behavior of measures or functions in a geometrical and statistical fashion. It was first introduced by Mandelbrot in the context of turbulence [*Mandelbrot*, 1969, 1972], even if the term *multifractal* was successively proposed by *Frisch and Parisi* [1985]. Some geological and geophysical applications were described by *Agterberg* [1997], *Barton and La Pointe* [1995], and *Cheng* [1999] for studying the spatial distributions of mineral deposits and hydrocarbon accumulations. The basic concept of multifractal analysis is to assess fractal dimensions of self-similar structures with varying regularities and to produce the distribution of indices of regularity, which constitutes the multifractal spectrum (MFS). Multifractal formalism relates the MFS to the partition function measuring high-order dependencies in the data. In recent years, multifractal formalism has been implemented with wavelets [*Bacry et al.*, 1996; *Arneodo et al.*, 1995; *Gonçalvès et al.*, 1998; *Riedi*, 1998, 1999; *Wendt et al.*, 2009]. This approach is very amenable to computation and estimation in practice. The advantages of using the wavelet-based MFS are availability of fast algorithms for wavelet transform, the locality of wavelet representations in both time and scale, and intrinsic dyadic self-similarity of basis functions. Wavelet-based fractal and multifractal spectra have been successfully applied in medical applications by *Nicolis et al.* [2011], *Ramírez-Cobo and Vidakovic* [2013], *Jeon et al.* [2014], and *Hermann et al.* [2015] for discriminating mammographic images.

In the following section we will briefly describe the basic concept of the 2-D wavelet transform and the wavelet-based fractal and multifractal spectrum (WMFS) (sections 2.2 and 2.3). Then we will apply the methods to a series of Landsat 8 images for assessing the presence of minerals or shale gas reservoirs (section 2.4). Some conclusions are provided in section 2.5.

2.2. 2-D Wavelet Transforms

Wavelets are the building blocks of wavelet transforms in the same way that the functions einx are the building blocks of the ordinary Fourier transform. But in contrast to sines and cosines, wavelets can be supported on an arbitrarily small closed interval. Basics on wavelets can be found in many texts, monographs, and papers (see, e.g., *Daubechies* [1992] or *Vidakovic* [1999], among others).

In two or higher dimensions wavelets provide an appropriate tool for analyzing self-similar objects. The energy preservation in orthogonal wavelet analysis, using Parseval's Theorem (see *Oppenheim and Schafer* [1975]), allows for defining wavelet spectra in a manner similar to that in the Fourier domains. Operationally the traditional tensor-product, 2-D wavelet transforms are constructed through the translations and the dyadic scaling of a product of univariate wavelets and scaling functions,

$$\phi\left(u_x,u_y\right) = \phi\left(u_x\right)\phi\left(u_y\right)$$
$$\psi^h\left(u_x,u_y\right) = \phi\left(u_x\right)\psi\left(u_y\right)$$
$$\psi^v\left(u_x,u_y\right) = \psi\left(u_x\right)\phi\left(u_y\right)$$
$$\psi^d\left(u_x,u_y\right) = \psi\left(u_x\right)\psi\left(u_y\right),$$

(2.1)

which are known as separable 2-D wavelets. The symbols *h, v,* and *d* in (1) stand for horizontal, vertical, and diagonal directions, respectively. The atoms capture image features in the corresponding directions (see *Nicolis et al.* [2011] and *Jeon et al.* [2014]).

Any function $f \in L_2(\mathbb{R}^2)$ can be represented as

$$f\left(\boldsymbol{u}\right) = \sum_k c_{j_0,k}\phi_{j_0,k} + \sum_{j>j_0}\sum_k\sum_i d^i_{j,k}\psi^i_{j,k}\left(\boldsymbol{u}\right),$$

(2.2)

where $\boldsymbol{u} = \left(u_x,u_y\right) \in \mathbb{R}^2, i \in \{h,v,d\}, \boldsymbol{k} = \left(k_1,k_2\right) \in \mathbb{Z}^2$, and

$$\phi_{j,k}\left(\boldsymbol{u}\right) = 2^j\phi\left(2^j u_x - k_1, 2^j u_y - k_2\right)$$
$$\psi^i_{j,k}\left(\boldsymbol{u}\right) = 2^j\psi^i\left(2^j u_x - k_1, 2^j u_y - k_2\right)$$

for $i = h,v,d$. $c_{j_0,k}$ and $d^i_{j,k}$ are scaling and wavelet coefficients, defined as

$$c_{j_0,k} = \int f\left(\boldsymbol{u}\right)\phi_{j,k}\left(\boldsymbol{u}\right)d\boldsymbol{u}$$

(2.3)

and

$$d^i_{j,k} = \int f\left(\boldsymbol{u}\right)\psi^i_{j,k}\left(\boldsymbol{u}\right)d\boldsymbol{u},$$

(2.4)

respectively [*Nicolis et al.*, 2011].

Wavelet coefficients can be computed by Mallat's algorithm [*Mallat*, 1998],

$$c_{j-1,l} = \sum_k h_{k-2l}c_{j,k}$$

(2.5)

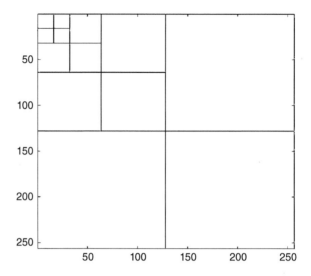

Figure 2.1 Tessellations for some 2-D wavelet transforms.

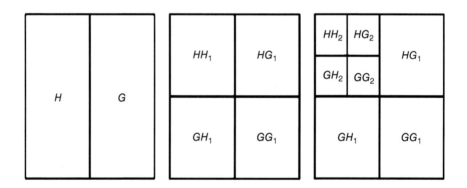

Figure 2.2 Steps of a two-level discrete wavelet transform of an image: wavelet transform on all rows (left); one-level decomposition (center); two-level decomposition (right).

and

$$d_{j-1,l} = \sum_{k} g_{k-2l} c_{j,k}, \qquad (2.6)$$

where h is the low-pass filter and g is the high-pass filter.

The result of a 2-D wavelet transform of an image is a square with squares-within-squares (Figure 2.1) of low-pass (**H**) operations and high-pass operations (**G**), as shown in Figure 2.2. In particular, the first step of the DWT performs the transform on all rows producing two subsets of coefficients as in Figure 2.1 (left): the left side of the matrix contains downsampled low-pass coefficients of each row and the right contains the high-pass coefficients. The application of the DWT on the columns of the matrix in Figure 2.2 (left) decomposes an image into a lower resolution approximation image (HH_1) as well as horizontal (GH_1), vertical (HG_1), and diagonal (GG_1) detail components as in Figure 2.2 (center). To compute a two-level decomposition, the DWT algorithm is again applied on the HH_1, which further decomposes the HH_1 part in four subbands, HH_2, GH_2, HG_2, and GG_2 (Figure 2.2, right). To produce coefficients that can reconstruct images with increased smoothness, one may repeat the process many times.

For example, if we consider the image in Figure 2.3 (top), the wavelet coefficients at different levels of resolution can be represented as in Figure 2.3 (bottom).

2.2.1. Wavelet Spectra of Self-Similar Processes

Wavelets and wavelet-based spectra have been instrumental in the analysis of self-similarity [*Flandrin*, 1989, 1992; *Wornell*, 1995]. For self-similar processes, the fractal dimension D is related to the Hurst exponent H through the linear relation

$$D + H = n + 1, \tag{2.7}$$

where n is the dimensional space (see *Mandelbrot* [1983]). Methods for estimating the Hurst exponent can be used for assessing fractal properties. A typical stochastic model for fractals is the fractional Brownian motion $B_H(u)$, which is a self-similar process of order $H \in (0,1)$ (see Figure 2.4).

For this process the wavelet coefficients are given by

$$d^i_{j,k} = 2^j \int B_H(u) \psi^i (2^j u - k) du, \tag{2.8}$$

where the integral is taken over $\mathbb{R}^2$ and $i = h$, v, or d. The detail coefficients are random variables with zero mean and variance

$$E\left[|d^i_{j,k}|^2 \right] = 2^{2j} \int\int \psi^i (2^j u - k) \psi^i (2^j v - k) \left[B_H(u) B_H(v) \right] du \tag{2.9}$$

[*Heneghan et al.*, 1996]. From (9) one can derive

$$E\left[|d^i_{j,k}|^2 \right] = \frac{\sigma_H^2}{2} V_{\psi^i} 2^{-(2H+2)j}, \tag{2.10}$$

where V_{ψ^i} depends only on wavelets ψ^i and exponent H, but not on the scale j (see *Nicolis* [2011] for the derivation of this result).

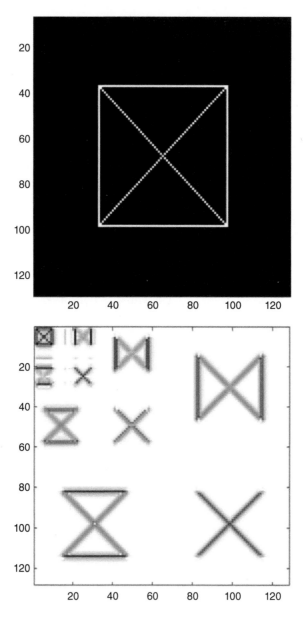

Figure 2.3 Image within a cross in a box with a cross within (top) and its 2-D wavelet transform using COIFLET2 basis and L = 3 levels of resolution.

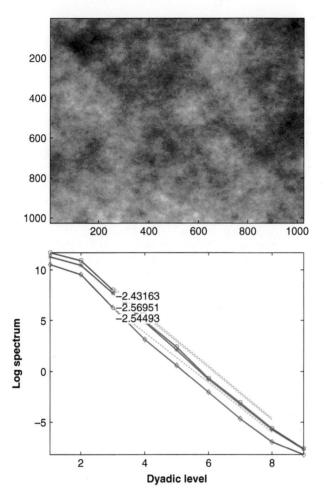

Figure 2.4 Simulated fractional Brownian motion with H = 0.33 (top) and its wavelet-based spectra using the wavelet Symmlet 6 (bottom). The Hurst coefficients in the diagonal, horizontal, and vertical directions, estimated by the slopes of the linear regressions using the relation –(2H+2), are given by Hd = 0.215, Hh = 0.28475, and Hv = 0.27247.

An application of the logarithm to both sides of Eq. (10) leads to the following linear relationship:

$$log_2 E\left[\left|d_{j,k}^i\right|^2\right] = -\left(2H+2\right)j + C_i,$$

(2.11)

where $C_i = log_2 \dfrac{\sigma_H^2}{2} V_{\psi^i}\left(H\right).$

The Hurst coefficient of an fBf is estimated from the slope of the linear equations given in (11).

2.3. Multifractal Formalism

Multifractal formalism is based on the concepts of the partition function and the Legendre transform (see *Chhabra and Jensen* [1989], *Gonçalvès et al.* [1998]). The partition function, $T(q)$, can be defined in terms of 2-D wavelet coefficients as

$$T^i(q) = \lim_{j \to -\infty} \log_2 E\left[\left|d^i_{j,k}\right|^q\right],$$ (2.12)

where $d^i_{j,k}$ are the wavelet coefficients at level j and location k, and q is the order of moments. We emphasize that q is a real number within a certain range covering the negative numbers as well [*Gonçalvès et al.*, 1998].

Even though Eq. (12) is very informative, the singularity measure is not explicit. *Gonçalvès et al.* [1998] proposed that the local singularity strength could be measured in terms of wavelet coefficients as

$$\alpha^i(\mathbf{u}) = \lim_{k2^j \to u} \frac{1}{j} \log_2 \left|d^i_{j,k}\right|^q,$$ (2.13)

where $k2^j \to \mathbf{u}$ means that $\mathbf{u} = (\mathbf{u}_x, \mathbf{u}_y) \in [2^{-j}k_1, 2^{-j}(k_1+1)] \times [2^{-j}k_{21}, 2^{-j}(k_2+1)]$ for $k = (k_1, k_2)$, and $j \to \infty$.

Smaller $\alpha^i(\mathbf{u})$ reflect more irregular behavior at $\mathbf{u}$ and thus more singularity at time $\mathbf{u}$. The index i in (13) corresponds to one of three directions in detail spaces of 2-D wavelet transform, horizontal, vertical, and diagonal. Any inhomogeneous process has a collection of local singularity strength measures, and their distribution $f(\alpha)$ forms the MFS. A useful and efficient tool to estimate the MFS is the Legendre transform. The Legendre transform of the partition function is defined as

$$f^i_L(\alpha) = \inf_q \left\{q\alpha - T^i(q)\right\}.$$ (2.14)

It can be shown that $f^i_L(\alpha)$ converges to the true MFS using the theory of large deviations [*Ellis*, 1984].

If we rearrange Eq. (12), it becomes

$$E\left[\left|d^i_{j,k}\right|^q\right] \sim 2^{jT^i(q)} \quad \text{as} \quad j \to -\infty.$$ (2.15)

A standard linear regression can be used to estimate the partition function $T^i(q)$ since the values $\left|d_{j,k}^i\right|^q$ could be easily obtained by moment-matching method. Formally,

$$\log_2 S_j^i(q) = jT^i(q) + \varepsilon_j, \qquad (2.16)$$

where $S_j^i(q) = \dfrac{1}{2^{2j}}\sum_{k_1=0}^{2^j-1}\sum_{k_2=0}^{2^j-1}\left|d_{j,k_1,k_2}^i\right|^q$ is the empirical q^{th} moment of the wavelet coefficients and the error term ε_j is introduced from the moment matching method when replacing the true moments with the empirical ones. The ordinary least square (OLS) estimator gives the estimation of the partition function,

$$T^i(q) = \sum_{j=j_1}^{j_2} a_j \log_2 S_j^i(q), \qquad (2.17)$$

where the regression weights a_j must verify the two conditions $\sum_j a_j = 0$ and $\sum_j ja_j = 1$ [Delbeke and Abry, 1998]). Thus, we can estimate $f(\alpha)$ through a local slope of $T^i(q)$ at values

$$\alpha^i(q_l) = \left[T^i(q_{l+1}) - T^i(q_l)/q_0\right], \qquad q_l = lq_0 \qquad (2.18)$$

as $f^i[\alpha^i(q_l)] = q_l\alpha^i(q_l) - T^i(q_l)$.

Multifractal spectra can even be found for monofractal processes, where the spectra generated from such processes are ramplike with a dominant (modal) irregularity corresponding to the theoretical Hurst exponent [Shi et al., 2006]. The MFS can be easily generalized to higher dimensions (see Derado et al. [2008] and Ramírez-Cobo and Vidakovic [2013]).

2.3.1. Multifractal Descriptors

Rather than operating with multifractal spectra as functions (densities), we summarize them by a small number of meaningful descriptors that have interpretation in terms of location and deviation from monofractality.

In practice, the multifractal spectrum can be approximately described by three canonical descriptors: (1) spectral mode (Hurst exponent, SM), (2) left slope (LS) or left tangent (LT), and (3) width spread (broadness, B) or right slope (RS) or right tangent (RT). A typical multifractal spectrum can be quantitatively described as shown in Figure 2.5 (bottom panel). In particular, SM represents the apex of spectrum or most common Hölder regularity index α found within the signal, and LS (or LT) represents the slope of the distribution produced by the collection of Hölder regularity index α with smaller values of the mode (SM). However, broadness (B) is a more intricate descriptor of the multifractal spectrum. B is believed to be more meaningful than RS or RT because it

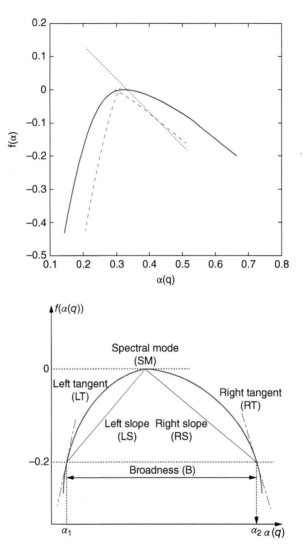

Figure 2.5 Top: Multifractal spectra for mono- (dashed line) and multifractal (solid line) processes (the dotted line is the theoretical slope of the spectrum for a monofractal process). Bottom: Illustration of geometric descriptors of multifractal spectra. Note that the horizontal axis represents values of Hölder regularity index $\alpha(q)$, while the vertical axis represents values proportional to the relative frequency of these indices, $f(\alpha(q))$.

is a compound measure representing the overall nature of the multifractal spectra, taking into account the overall variability among the Hölder regularity index α. In addition, B partially accounts for right slope RS or right tangent RT in calculation, as the resultant value of B is based on the relative values of RS

and *LS*. Both slopes (or both tangents) can be obtained easily using the interpolation technique, while it is not straightforward to define *B* automatically. There are many ways to define *B*) In this work, we select the method proposed by [*Shi et al.*, 2006]. The overall multifractal descriptors are also graphically presented in Figure 2.5 (bottom).

2.4. Applications to Landsat 8 Images

In this section, we consider three groups of Landsat 8 images downloaded from the web site http://earthexplorer.usgs.gov/. The first group of images represents an area in the north of Chile where there is a large concentration of minerals such as copper and gold. The second group of images shows a large shale gas reserve in the United States, and the third group is constituted by two Landsat images covering the Vaca Muerta region in Argentina. For all images we only worked with band 4 since it has been shown that it clearly reflects outcropping rocks, including alteration zones associated with gold and copper mineralization in the area (see *Cheng* [1999]).

2.4.1. Multifractal Analysis for Assessing Mineral Deposits

Multifractal analysis of fractures and mineral deposits has been largely explored using some known methods such as the multifractal detrended analysis or extensions of the box-counting method. The wavelet-based multifractal spectrum constitutes a new tool for studying the local singularities in signals or images. Even if the method has shown it is powerful in exploring fractal and multifractal features in many fields of science (medicine, turbulence, clouds modelling, etc.), there are not applications in the context of mineral characterization using satellite images. In this section, we try to find some evidences that the wavelet-based multifractal spectrum can be used for detecting the presence of mineral deposits in a given area. To this end, we use five satellite images covering the north part of Chile as represented in Figure 2.6. Some of these images are highly characterized by the presence of copper (represented by the red circles in Figure 2.6a) and gold (represented by the blue circles in Figure 2.6a).

In order to apply the wavelet based multifractal algorithm, we transformed the images in numerical matrices of size 7021×7991 using the MatLab program. Since the wavelet method works on images of size $2^j \times 2^j$ we split each image in squares of size 1024×1024 as in Figure 2.7.

Figure 2.8 is an example of the application of the multifractal spectra: first we represented the area of study as in Figure 2.8a and 2.8b, then we used the wavelet Symmlet 8 for evaluating the wavelet-based multifractal spectrum. By repeating the process for each square of each figure, we obtained a number of 125 for each descriptor (given by H = Hurst exponent; R = right slope, L = left slope, and B = broadness).

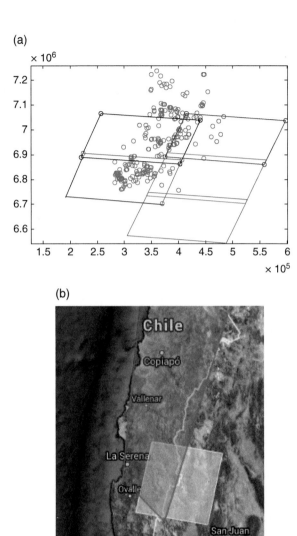

Figure 2.6 (a) Representation of the areas covered by the Landsat 8 images using UTM coordinates. The red circles represent the presence of a copper mine and the blue ones the presence of a gold mine. (b) The white quadrilateral represents the Landsat 8 taken at 7:20 a.m. on 16 April 2014 and delimited by the coordinates (−29.3792; −70.6048), (−29.6530; −68.72236), (−31.2380−69.1236), and (−30.9599 −71.0535). (c) The light blue quadrilateral represents the Landsat 8 taken at 7:25 a.m. on 16 April 2014 delimited by the coordinates (−27.9415; −70.2240), (−28.2116; −68.3665), (−29.7973; −68.7602), and (−29.523; −70.6450). (d) The violet quadrilateral represents the Landsat 8 taken at 5:37 a.m. on 13 October 2014 and delimited by the coordinates (−26.5198; −69.8553), (−26.7865 −68.0209), (−28.3729; −68.408), and (−28.1024; −70.2677). (e) The green quadrilateral represents the Landsat 8 taken at 3:11 a.m. on 16 April 2014 and delimited by the coordinates (−27.9380; 71.8074), (−28.2081; −69.9500), (−29.7938; −70.3437), and (−29.5197; −72.2284). (f) The light–green quadrilateral represents the Landsat 8 taken at 3:10 a.m. on 16 April 2014 and delimited by the coordinates (−26.509; −71.4366), (−26.7753; −69.6024), (−28.3617; −69.9892), and (−28.0912; −71.8489).

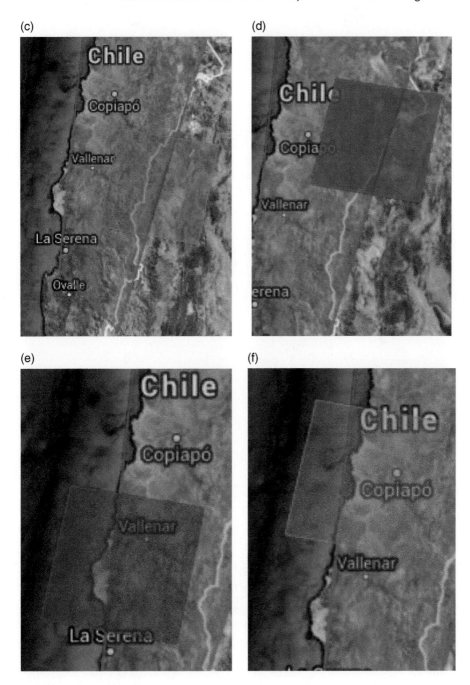

Figure 2.6 (Continued)

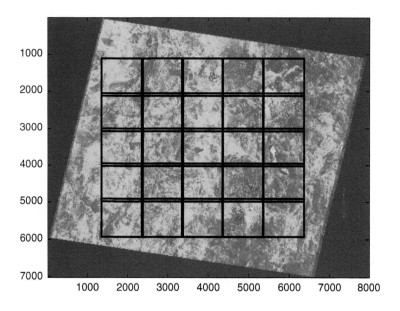

Figure 2.7 Landsat 8 image represented in Figure 2.6d. Black overlaying squares show the areas used for the application of the wavelet multifractal spectrum.

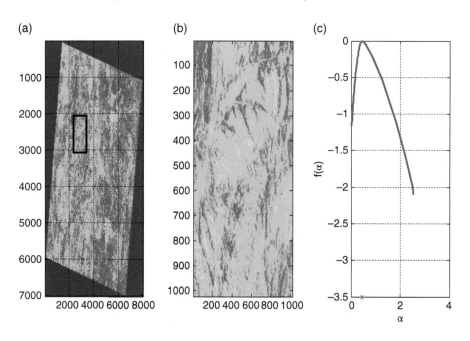

Figure 2.8 Wavelet-based multifractal analysis of Landsat 8 image of Figure 2.6d: (a) The black rectangle shows the selected area which has been zoomed in (b). (c) Wavelet multifractal spectrum using the wavelet Symmelet 8.

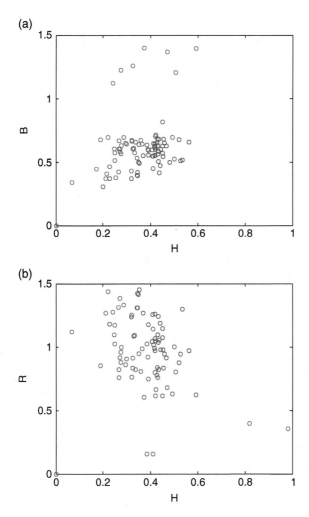

Figure 2.9 (a) Estimates of H vs. B values for all squares. (b) Estimates of H vs. R values for all squares. Red circles represent the descriptors evaluated in the squares where there are not mines, and blue circles show the descriptors in the areas where there is almost a mine of copper or gold.

From Figure 2.9, we can note that the multifractal descriptors H, B, and R can discriminate toward areas with the presence of minerals. In particular, it seems that areas with mineral deposits are in general characterized by lower values of B (that is, the broadness of the wavelet spectrum is smaller) and higher values of H (except for some atypical values).

2.4.2. Multifractal Analysis of Landsat 8 Images of a Shale Gas Reservoir in the United States

The aim of this application is to see if the shale gas, normally detected at high depth, might present some geological anomalies that could be detected by the multifractal analysis using satellite images. We collected eight different Landsat 8 images covering the shale gas reservoir in a zone of the United States including parts of Wyoming, Utah, and Colorado. Areas of shale gas reservoir and satellite images are represented in Figure 2.10.

From Figure 2.11b it is evident that the superficial rocks present a fractal structure. We then apply the fractal and multifractal spectrum for exploring the mono/multifractal features. Since the fractal spectra were approximately similar for each direction (vertical, horizontal, and diagonal), we decided to evaluate the multifractal spectrum only in the diagonal direction.

The nonlinear behavior of the fractal spectra (see, e.g., Figure 2.12a) justifies the use of multifractal analysis for this application (Figure 2.13).

The main drawback of using Landsat 8 images for fractal and multifractal analysis is the different resolution of each image (see, e.g., Figure 2.10c where the excessive smoothness of the image, due to low resolution, produces larger Hurst coefficients). Also, the different resolution produces a misclassification of the multifractal descriptors. In Figure 2.14 the blue and green circles represent the pairs (H,B) (Figure 2.14a) and (H,R) (Figure 2.14b) using three low-resolution images correspondent to the quadrants of Figure 2.10 b, c, and d.

In Figure 2.15 we plotted the pairs of descriptors (H,B) (Figure 2.15a) and (H,R) (Figure 2.15b) without considering the results from the low-resolution images of Figure 2.10 b, c, and d. In particular, we denoted by red circles the pairs of multifractal descriptors resulting from the satellite areas external to the shale gas reservoir and by blue circles the multifractal analysis of the quadrants within the shale gas reservoir area. Due to the low number of available points, it is difficult to assess the discriminatory power of the multifractal descriptors. Moreover, if we compare the pairs of the descriptors (H,B) of the shale gas reservoirs with those obtained in section 2.4.1 for the mineral deposits, we can note a significant difference: mineral deposits are in general characterized by higher values of H (for approximately equal values of B).

2.4.3. Application to the Shale Reservoir of Vaca Muerta, Argentina

In this section we consider the Landsat 8 images which include the shale gas deposit at Vaca Muerta, in the province of Neuquen, Argentina. Vaca Muerta constitutes the second-largest reserves of shale gas in the world. In particular, we consider two satellite images as represented in Figure 2.16. The first satellite

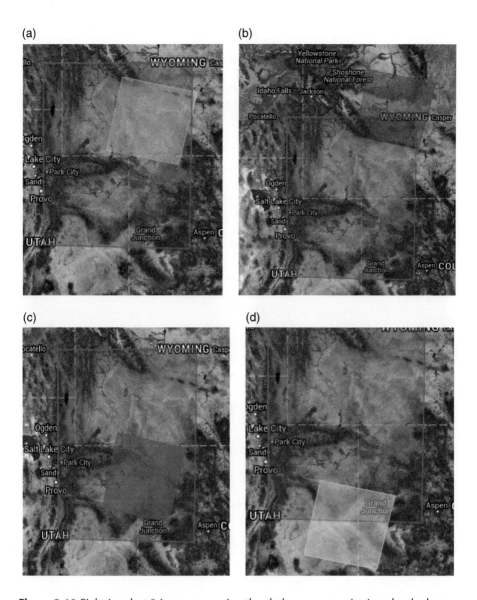

Figure 2.10 Eight Landsat 8 images covering the shale gas reservoir given by the brown large square. The first satellite image (a) was collected on 7 February 2015 at 9:11 a.m. with latitudes between 42.32 and 44.13, and longitudes between –109.30 and –106.32; the second (b), on 28 January 2015 at 1:25 p.m. with latitudes between 40.81 and 42.70, and longitudes between –109.74 and –107.03; the third (c) on 7 February 2015 at 3:53 p.m. with latitudes between 39.38 and 41.27, and longitudes between –110.17 and –107.53; the fourth (d) on 7 February 2015 at 3:51 p.m. with latitudes between 37.95 and 39.83, and longitudes between –110.59 and –108.01; the fifth (d) on 3 July 2014 at 3:08 a.m. with latitudes between 37.95 and 39.83, and longitudes between –112.16 and –109.57; the sixth (e) on 9 June 2014 at 5:04 p.m. with latitudes between 39.38 and 41.27, and longitudes between –111.72 and –109.10; the seventh (f) on 9 June 2014 at 7:34 p.m. with latitudes between 40.81 and 42.70, and longitudes between –111.31 and –108.61; and the eighth (g) on 7 February 2015 at 10:47 a.m. with latitudes between 37.96 and 39.84, and longitudes between –112.16 and –109.57.

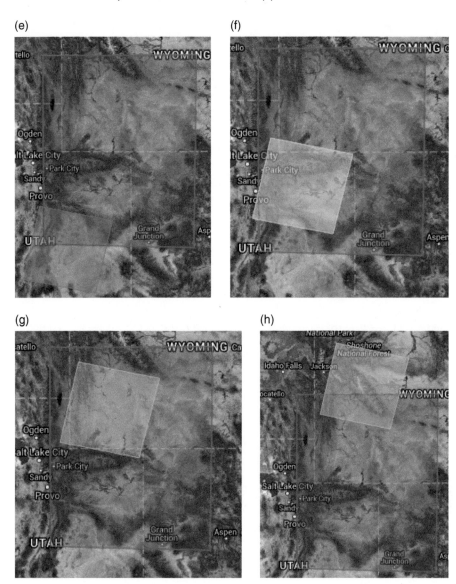

Figure 2.10 (Continued)

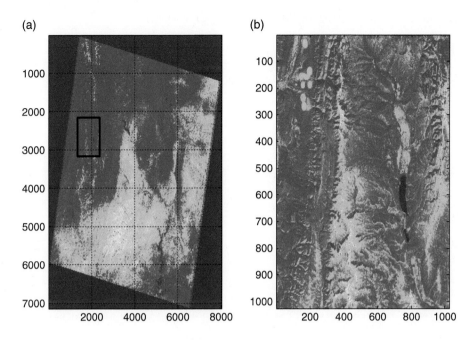

Figure 2.11 (a) Landsat 8 image represented in Figure 2.6d. Black rectangle represents the selected area for the fractal and multifractal analysis. (b) Zoom of the image in the black rectangle.

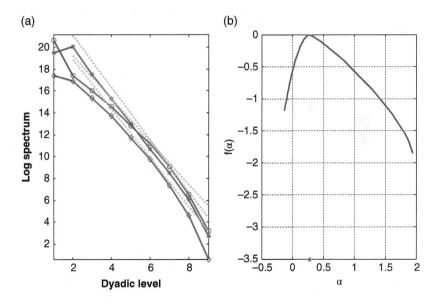

Figure 2.12 Fractal (a) and multifractal (b) spectra of Landsat 8 image of Figure 2.10(b) using the wavelet Symmlet 8.

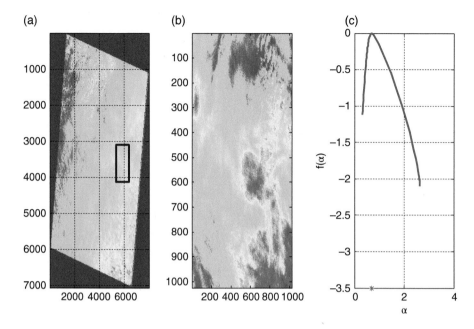

Figure 2.13 (a) LANDSAT-8 images represented in Figure 2.6 (d). The black rectangle shows the area used for the application of the wavelet multi fractal spectrum. (b) Zoom of the area included in the black rectangle and (c) its multi fractal spectrum.

image, delimited by the coordinates (–36.4830; –71.7382), (–36.5143; –72.189) (–37.9547; –69.4804), (–39.8315; –72.2766;), and (–39.8768; –69.49349) covers the north part of the shale gas reservoir (Figure 2.16c), and the second image, delimited by the coordinates (–37.9124; –72.1895), (–37.9547; –69.4804), (–39.8315; –72.2766), and (–39.8768; –69.4935), covers the south part (Figure 2.16d).

As in the applications described in sections 2.4.1 and 2.4.2, we divided each satellite image into 25 subimages where we evaluated the wavelet transform and determined the multifractal spectra for each area. Figure 2.17 shows an example of a selected area in the northern image.

All images show a multifractal structure with the major number of estimated Hurst coefficientes between 0.3 and 0.55, with broadness values between 0.4 and 0.7, and right slope values between 0.8 and 2.2 (see Figure 2.18). Moreover, the number of multifractal descriptors for the area of Vaca Muerta reservoir are not sufficient for an adequate discriminatory analisis.

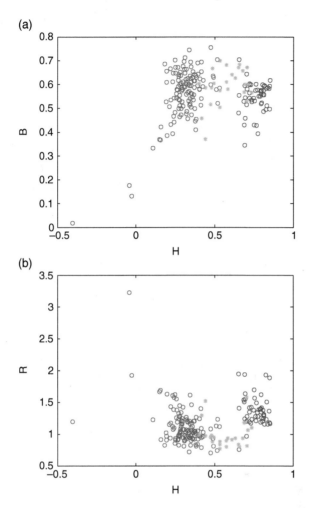

Figure 2.14 Classification of multifractal descriptors for all images: (a) Red circles are the pairs (B,H) resulting from the multifractal analysis of the quadrants in Figure 2.10 a, e, f, g, and h; Green circles are the pairs (B,H) resulting from the multifractal analysis of the quadrants in Figure 2.10b; and blue circles are the pairs (B,H) resulting from the multifractal analysis of the quadrants in Figure 2.10c and d. (b) The same classification of colors has been considered for the pairs of descriptors (H,R).

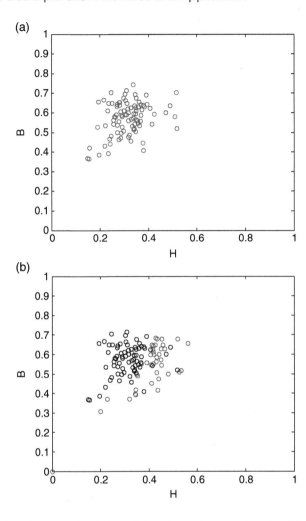

Figure 2.15 Classification of multifractal descriptors for some images: (a) Red circles are the pairs (H,B) resulting from the areas external to the shale gas reservoir (Figure 2.10b and h), while blue circles are the pairs (H,B) resulting from the multifractal analysis of the quadrants within the shale gas reservoir area (Figure 2.10a, and g). (b) Comparison between shale gas reservoir and mineral deposits: black circles are the pairs (H,R) resulting from the areas within the shale gas reservoir, while blue circles are the pairs (H,R) resulting from the multifractal analysis of areas including mineral deposits in Chile (see Section 4.1).

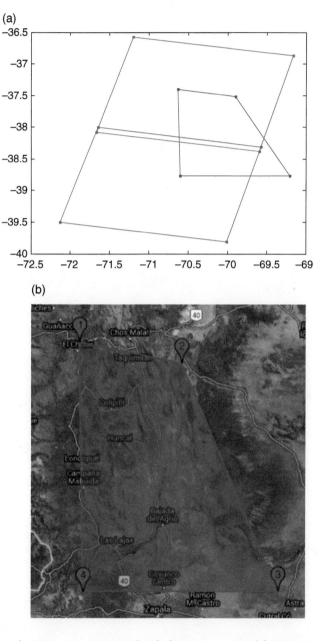

Figure 2.16 Landsat 8 images covering the shale gas reservoir of the Vaca Muerta region in Argentina. (a) The blue quadrilateral shows the area of the Vaca Muerta reservoir; the green and red quadrilaterals indicate the areas for the north and south satellite images, respectively. (b) Shale gas reservoir area of Vaca Muerta Region. (c) North Landsat 8 image taken at 6:18 p.m. on 20 October 2014. (d) South Landsat 8 image taken at 6:10 p.m. on 20 October 2014.

(c)

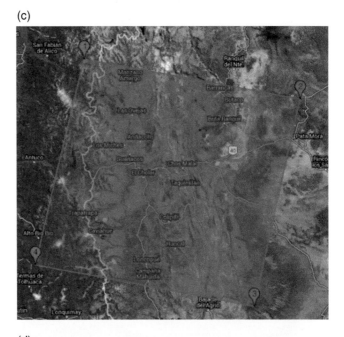

(d)

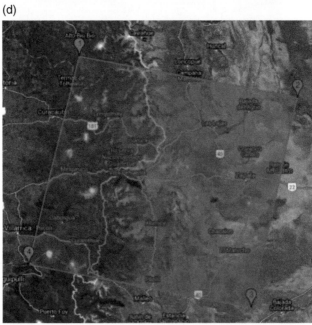

Figure 2.16 (Continued)

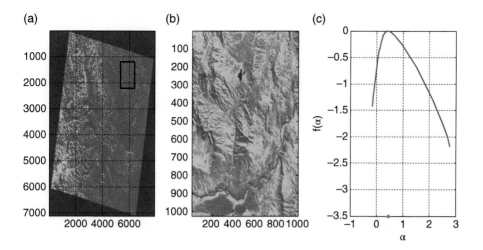

Figure 2.17 Wavelet-based multifractal analysis of northern Landsat 8 image: (a) The black rectangle shows the selected area of the Landsat 8 that covers the north part of the Vaca Muerta shale gas reservoir. (b) Zoom of the image selected in the black rectangle. (c) Multifractal spectrum using the wavelet Symmelet 8.

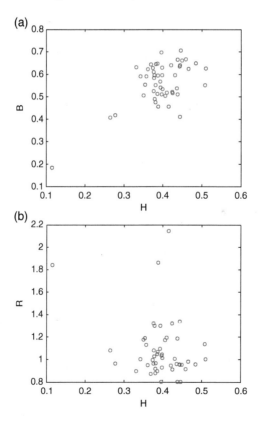

Figure 2.18 (a) The red circles are the pairs (H,B) resulting from the northern Landsat 8 image, and the blue circles show the pairs (H,B) resulting from the southern Landsat 8 image. (b) (a) Red circles are the pairs (H,R) resulting from the Northern Landsat 8 image and the blue circles show the pairs (H,R) resulting from the Southern Landsat 8 image.

2.5. Conclusions and Further Developments

The wavelet-based multifractal spectrum constitutes a new tool for exploring the presence of mineral deposits and shale gas reservoirs using Landsat 8 images. Although the method seems to work for assessing the presence of mineral deposits in some areas of Chile, we think that higher resolution images are necessary for testing the discriminatory performance of the multifractal descriptors in the case of shale gas reservoirs. The principal result of this analysis is that, in all cases, the superficial rocks have a multifractal structure for which descriptors might be used in a discriminatory analysis or to complement the results coming from geological analysis.

Acknowledgment

I thank Bruno Barra of ThinkInnova Spa for his precious help in collecting satellite images.

References

Agterberg, F. P. (1997), Multifractal modelling of the sizes and grades of giant and supergiant deposits. *Global Tectonics and Metallogeny*, 6 (2), 131–136.

Arneodo, A., E. Bacry, and J. Muzy (1995), The thermodynamics of fractals revisited with wavelets. *Physica A*, 213, 232–275.

Bacry E., A. Arneodo, J. Muzy, and P. Graves (1996), A wavelet based multifractal formalism: Application to the study of DNA sequences. *J. Tech. Phys.*, 37(3–4), 281–284.

Barberes G. (2015), Remote sensing (Landsat–8) applied to shale gas potential evaluation. The case of South Portuguese Zone. Proceedings of *European Regional Conference and Exhibition*, Lisbon, May 18–19.

Barton C. C., and P. R. La Pointe (1995) *Fractals in Petroleum Geology and Earth Processes*. Plenum Press, New York and London.

Barton C. C., C. H. Scholz, and P. R. H. Schutter, and W. J. Thomas (1991), Fractal nature of hydrocarbon deposits, 2: Spatial distribution. Abstract: *Am. Assoc. Petrol. Geol. Bull.*, 75(5), 53939.

Chen Z., K. Osadetz, P. Hannigan (2001), An improved fractal model for characterizing spatial distribution of undiscovered petroleum accumulations. *Annual Conference of the International Association for Mathematical Geology*, Cancun, Mexico.

Cheng, Q. (1999), Multifractality and spatial statistics. *Computers and Geosciences*, 25(9), 949–961.

Chhabra A., and R. Jensen (1989), Direct determination of the $f(\alpha)$ singularity spectrum, *Physical Review Letters*, 62(12), 1327–1330.

Crovelli, R. A., and C. Barton (1995), Fractal and the Pareto distribution applied to petroleum accumulation-size distribution. In *Fractals in Petroleum Geology and Earth Processes*, edited by C. Barton and P. R. La Pointe, Plenum Press, New York, 59–72.

Daubechies, I. (1992), *Ten Lectures on Wavelets*. CBMS-NSF Regional Conference Series in Applied Mathematics, Philadelphia, Pennsylvania 19104-2688: Society for Industrial and Applied Mathematics.

Delbeke, L., and P. Abry (1998), Wavelet-based estimators for the self-similar parameter of fractional Brownian motion. Submitted to *Appl. Comp. Harm. Anal.*

Derado, G., K. Lee, O. Nicolis, F. D. Bowman, M. Newell, F. Ruggeri, and B. Vidakovic (2008), Wavelet-based 3-D multifractal spectrum with applications in breast MRI images. In *Bioinformatics Research and Applications* (Mandoiu, Sunderraman, and Zelikovsky, eds.), Lecture Notes in Computer Science, Springer, 281–292.

Drew L. J., E. D. Attanasi, and J. H. Schuenemeyer (1988), Observed oil and gas field distributions: A consequence of discovery process and prices of oil and gas, *Math. Geology*, 20(8), 939–953.

Ellis, R. S. (1984), Large deviations for a general class of random vectors, *The Annals of Probability*, 12(1), 1–12.

Flandrin, P. (1989), On the spectrum of fractional Brownian motions. *IEEE Transaction on Information Theory*, 35, 197–199.

Flandrin, P. (1992), Wavelet analysis and synthesis of fractional Brownian motion. *IEEE Transaction on Information Theory*, 38, 910–917.

Frish, U., and G. Parisi (1985), On the singularity structure of fully developed turbulence. In *Turbulence and Predictability in Geophysical Fluid Dynamics and Climate Dynamics*, edited by M. Gil, R. Benzi, and G. Parisi, pp. 84–88, Elsevier, Amsterdam.

Gonçalvès P., H. Riedi, and R. Baraniuk (1998), Simple statistical analysis of wavelet based multifractal spectrum estimation. In *Proceedings 32nd Asilomar Conference on Signals, Systems and Computers*, Pacific Grove, CA.

Heneghan C., S. B. Lown, and M. C. Teich (1996), Two dimensional fractional Brownian motion: Wavelet analysis and synthesis. *Image Analysis and Interpretation, Proceedings of the IEEE Southwest Symposium*, 213–217.

Hermann P., T. Mrkvicka, T. Mattfeldt, M. Minrov, K. Helisov, O. Nicolis, F. Wartner, and M. Stehlk (2015), Fractal and stochastic geometry inference for breast cancer: Case study with random fractals models and quermass-interaction process, *Statistics in Medicine*. DOI: 10.1002sim.6497.

Houghton, J. C. (1988), Use of the truncated shifted Pareto distribution in assessing size distributions of oil and gas fields, *Mathematical Geology*, 20(8), 907–938.

Jeon S., O. Nicolis, and B. Vidakovic (2014), Mammogram diagnostics via 2-D complex wavelet based self-similarity measures, *Sao Paulo Journal of Mathematical Sciences*, 8(2), 265–284.

La Pointe, P. R. (1995), Estimation of undiscovered hydrocarbon potential through fractal geometry. In *Fractal in Geology and Earth Sciences*, edited by C. C. Barton and P. R. La Pointe, pp. 35–57, Plenum Press, New York.

Mallat, S. G. (1998), *A Wavelet Tour of Signal Processing*, Academic Press, New York.

Mandelbrot, B. B. (1962), Statistics of natural resources and the law of Pareto, IBM Research Note NC-146.

Mandelbrot, B. (1969), Long-run linearity, locally Gaussian process, H-spectra, and infinite variances, *International Economic Review*, 10, 82–111.

Mandelbrot, B. (1972), A statistical methodology for non-periodic cycles: From the covariance to R/S analysis, *Annals of Economic and Social Measurement*, 1, 259–290.

Mandelbrot, B. (1983) *The Fractal Geometry of Nature*, Macmillan, New York.

Nicolis O., P. Ramirez, and B. Vidakovic (2011), 2-D wavelet-based spectra with applications, *Computational Statistics and Data Analysis*, 55(1), 738–751.

Oppenheim, A. V., and R. W. Schafer (1975), *Digital Signal Processing*. Prentice Hall.

Ramírez-Cobo, P., and B. Vidakovic (2013), A 2D wavelet-based multiscale approach with applications to the analysis of digital mammograms, *Computational Statistics & Data Analysis*, 58, 71–81.

Riedi, R. H. (1999,) Multifractal processes. *Technical Report,* 99-06.

Riedi, R. H. (1998), Multifractals and wavelets: A potential tool in geophysics. *Proceedings of the SEG Meeting*, New Orleans, Louisiana.

Shi, B., K. P. Moloney, Y. Pan, V. K. Leonard, B. Vidakovic, J. A. Jacko, and F. Sainfort (2006), Wavelet classification of high frequency pupillary responses. *Journal of Statistical Computation and Simulation*, 76(5), 431–445.

Vidakovic, B. (1999), *Statistical Modeling by Wavelets*, John Wiley and Sons, New York.

Wendt, H., et al. (2009), Wavelet leader multifractal analysis for texture classification. *Proc. IEEE Int. Conf. Image Proc. (ICIP)*, 3829–3832.

Wornell, G. (1995), *Signal Processing with Fractals: A Wavelet Based Approach*, Prentice Hall.

3

3. SEISMIC SIGNAL DENOISING USING EMPIRICAL MODE DECOMPOSITION

Said Gaci

Abstract

Denoising is a critical step in the analysis of geophysical data. This study aims to suggest a denoising method using empirical mode decomposition (EMD). The proposed technique has been compared with the discrete wavelet transform (DWT) thresholding. First, both methods have been applied on simulated signals with different waveforms ("blocks," "heavy sine," "Doppler," and "mishmash"). It is shown that the EMD denoising method is the most efficient for blocks, heavy sine, and mishmash signals for all the considered signal-to-noise ratio (SNR) values. However, for the Doppler signal, the DWT thresholding yields the best results, and the difference between the calculated mean square error values using the two studied methods is small and decreases as the SNR values get smaller. Moreover, both denoising techniques have been implemented on real seismic traces recorded in the Algerian Sahara. The best denoising results have been derived using the suggested technique. In addition, it is worth noting that, unlike the DWT-based method that needs to choose the optimal wavelet thresholding parameters, the suggested technique offers the advantage of not requiring any preselected parameter. To conclude, it is demonstrated that the proposed method can serve as a good tool for denoising signals.

Sonatrach- IAP, Boumerdès, Algeria

Oil and Gas Exploration: Methods and Application, Monograph Number 72,
First Edition. Edited by Said Gaci and Olga Hachay.
© 2017 American Geophysical Union. Published 2017 by John Wiley & Sons, Inc.

3.1. Introduction

Denoising of signals is a classic issue in signal processing. Noise is considered as any undesired signal that interferes with the wanted information signal. The objective of any signal denoising method is to efficiently lessen the noise level and to highlight the useful signal, with a minimum of information loss.

Time-frequency analysis methods, such as short-time Fourier transform or wavelets are considered more suitable to process nonstationary signals. Therefore, they are often used for real signal denoising in either the time or frequency domain.

Despite achieving a suitable time-frequency resolution, wavelets require a specified basis function, the choice of which is sometimes difficult. Recently, empirical mode decomposition (EMD) has been suggested by N. Huang [*Huang et al.*, 1998]. It has proven to be a powerful tool to analyze nonstationary and nonlinear signals. It is completely adaptive and data driven, without needing a priori basis function selection for signal decomposition.

EMD has been successfully used for signal denoising in a wide range of applications, such as biomedical signals [*Weng et al.*, 2006], acoustic signals [*Khaldi et al.*, 2008], and ionosphere signals [*Tsolis and Xenos*, 2009].

In this study we propose an EMD-based method for Gaussian noise removal. This technique is compared with the discrete wavelet transform (DWT) thresholding through application on synthetic and real data. The remainder of the present chapter is structured as follows. First, a brief mathematical background of both denoising methods is given. Then, we present the results obtained by these techniques from different synthetic signal types and real seismograms recorded in the Algerian Sahara.

3.2. Theory

3.2.1. EMD Algorithm

EMD is a signal decomposition allowing a nonlinear and nonstationary signal to be separated into a set of intrinsic mode functions (IMFs) plus a residual function without requiring an a priori basis, as is the case with the Fourier and wavelet-based methods [*Huang et al.*, 1998, 1999; Flandrin and Gonçalvès, 2004]. Each IMF corresponds to a monocomponent signal or an oscillatory mode with one instantaneous frequency. An IMF must satisfy two basic criteria: (a) The number of the extreme and the number of zero crossings must be equal or must differ by one at most. (b) The mean of the envelopes determined by the local maxima and minima, respectively, called the upper and lower envelope, should be zero.

The most important process of the EMD algorithm to extract IMFs is called sifting. It is an iterative procedure as follows:

1. Find local extreme (maxima and minima) of the signal $X(z)$.
2. Compute the interpolating signals: the upper $U(z)$ and lower $L(z)$ envelope of the signal from, respectively, the local maxima and minima by cubic spline interpolation.
3. Compute the mean of the local envelopes: $m(z) = \dfrac{U(z) + L(z)}{2}$
4. Extract the local mean $m(z)$ from the signal: $h_1(z) = X(z) - m(z)$
5. Replace the signal $X(z)$ with $h_1(z)$, and repeat steps 1–4 until the obtained signal satisfies the two IMF conditions.

The sifting process is terminated by using one of the following conditions: after extracting n IMFs, the residue, $r_n(z)$ is either an IMF or a monotonic function. More details on the stopping criteria can be found in previous researches [*Huang et al.*, 1998, 1999, 2003; *Huang*, 2005; *Flandrin and Gonçalvès*, 2004; *Rilling et al.*, 2003; *Rilling and Flandrin*, 2008].

Hence, the original signal can be reconstructed by the sum of IMFs described by

$$X(z) = \sum_{m=1}^{n-1} \mathrm{IMF}_m(z) + r_n(z), \tag{3.1}$$

where n-1 is the number of IMFs, i.e. the signal is decomposed into $(n$-$1)$ IMFs and one residual.

EMD offers an alternative approach for signal analysis such as instantaneous frequency, autoregressive parameter estimation, classification, and denoising, which is the most popular application of EMD.

3.2.2. Gaussian Noise Model and EMD Signal Denoising

Fractional Gaussian noise (fGn) is defined as the increment process of fractional Brownian motion. The statistical properties of fGn are determined by a scalar parameter, H, known as the Hurst exponent, and its autocorrelation function is given by

$$r(k) = \frac{\sigma^2}{2}\left(\left|k-1\right|^{2H} + \left|k\right|^{2H} - \left|k+1\right|^{2H}\right), \tag{3.2}$$

where the variance of the process is σ^2.

For the special case $H = 1/2$, the fGn process is reduced to a white noise, whereas other values correspond to colored Gaussian noise with non-zero correlations, negative when $0 < H < 1/2$ or positive when $1/2 < H < 1$ (long-range dependence). Simulations with fGn and white noise showed that EMD behaves as a dyadic filter bank [*Flandrin et al.*, 2005]. In addition, the log-variance of the IMFs varies linearly with the Hurst parameter (*H*) [*Wu and Huang*, 2004; *Flandrin et al.*, 2004]:

$$\log_2 V_H(i) = \log_2 V_H(2) + 2(H-1)(i-2)\log_2 \rho_H. \tag{3.3}$$

For $i \geq 2$ and $\rho_H = 2.01$, $V_H(i)$ is the variance of the *i*th IMF. Then the energy of each of the IMFs can be given as a function of the first IMF energy (E_1):

$$E_k = \left(E_1 \, 2.01^{-k}\right)/0.719, \quad k = 2,3,\ldots,n-1, \tag{3.4}$$

where E_1 can be approximated by the variance of the first IMF of the noisy signal.

A simple denoising approach based on EMD is suggested by *Flandrin et al.* [2005], using the above model. It consists of decomposing the noisy signal into IMFs, and then comparing the IMF energies with the theoretical estimated noise-only IMF energies calculated using relation (3.4). Finally, the signal reconstruction is performed by summing the IMFs, whose energy significantly deviates from the theoretical noise model. In order to discriminate between IMFs originating from the noisy signal and the noise-only one, we set a difference threshold for $log_2 E_k$. If the difference is less than the threshold, the corresponding IMF is considered a noise component and is rejected. Otherwise, it is kept in the signal reconstruction.

In this chapter, we empirically set the threshold as $|0.01 \, log_2 E_1|$,

$$\begin{cases} if \ \log_2(E(IMF_k)) - \log_2(E_k) \geq |0.01\log 2(E_1)|, & IMF_k \ is \ kept \\ Otherwise, & IMF_k \ is \ rejected \end{cases}, k = 2,3,\ldots,n-1$$

3.2.3. Discrete Wavelet Transform

A mother (or analyzing) wavelet ψ is a function with a zero mean and finite energy [*Daubechies*, 1988; *Goupillaud et al.*, 1984]. The wavelet family ψ_{ab} is derived by using the scale "a" and the location "b" parameters:

$$\psi_{ab}(t) = \frac{1}{\sqrt{a}} \psi\left(\frac{t-b}{a}\right) a > 0, b \in R. \tag{3.5}$$

The parameters "a" and "b" can be discretized as follows:

$$a_m = a_0^m, \ b_n = nb_0 a_0^m; \ m,n \in Z, \tag{3.6}$$

with $a_0 > 1$ being a fixed dilation coefficient and $b_0 > 0$ dependent on the mother wavelet ψ. The specific discretization choice, $a_0 = 2$ and $b_0 = 1$, leads to a set of orthogonal bases formed by the resulting wavelet family (given by eq. 3.5):

$$\psi_{mn}(t) = a_0^{-m/2} \psi \left(a_0^{-m} t - nb_0 \right). \tag{3.7}$$

In this orthonormal wavelet basis, all functions $f \in L^2(R)$ can be approximated by linear combinations [*Mallat*, 1989]:

$$f(t) = \sum_{m \in Z} \sum_{n \in Z} C_{mn} \psi_{mn}(t), \tag{3.8}$$

where C_{mn} is given by

$$C_{mn} = \int_{-\infty}^{+\infty} f(t) . \bar{\psi}_{mn}(t) . dt, \tag{3.9}$$

where the above-line symbol "–" denotes the complex conjugate. The coefficient C_{mn} measures the contribution of the scale 2^m in the position $n2^m$.

3.2.4. Wavelet Thresholding

Let the original signal be x_i, $i=0, N-1$, where N is its length. The signal has been corrupted by additive independent and identically distributed (i.i.d.) Gaussian random noise $z_i \propto N(0,\sigma^2)$ as follows:

$$y_i = x_i + z_i \ (i = 0,1,\ldots,N-1), \tag{3.10}$$

where σ is the standard deviation of the noise, also called the level of the noise. The goal is to eliminate the noise, or "denoise," i.e., to shrink the contribution of the noisy component, and thus to emphasize the signal x_i.

Mathematically, the DWT can be expressed as the product of the discrete samples of the signal and the orthonormal matrix W:

$$c = W.y \tag{3.11}$$

The denoising scheme based on the wavelet thresholding is given by

$$\tilde{x} = W^{-1}\left(T(Wy)\right), \tag{3.12}$$

where $T(\cdot)$ is a wavelet thresholding function. Indeed, wavelet denoising is a three-step process:
1. Compute the discrete wavelet transform of the noisy signal (Wy)
2. Pass it through the thresholding function $T(\cdot)$
3. Finally, apply the inverse wavelet transform to obtain the denoised signal ($\tilde{x}$)

In practice, two well-known types of $T(\cdot)$ are found: the hard-thresholding function (T_λ^h) and the soft-thresholding function (T_λ^s), which is also known as the wavelet shrinkage function.

$$T_\lambda^h = \begin{cases} c, & if\,|c| \geq \lambda \\ 0, & otherwise \end{cases} \qquad T_\lambda^s = \begin{cases} sign(c)(|c|-\lambda), & if\,|c| \geq \lambda \\ 0, & otherwise \end{cases} \tag{3.13}$$

The threshold λ is selected according to the signal energy and the noise level. The choice of the λ value is critical in denoising, since a small value yields an output comparable to the initial signal, whereas a high value can oversmooth the signal. In the present study, four threshold selection rules are used [*Donoho and Johnstone*, 1994; *Stein*, 1981; *Honório et al.*, 2012; *Gaci*, 2013]:
- "minimaxi": $\lambda = 0.3936 + 0.1829*(log\,(N)/log(2))$ where N is the length of the signal.
- "universal': $\lambda = \sigma\sqrt{2.\log N}$ where σ^2 is the noise variance and N is the length of the signal.
- "rigorous SURE" (rig sure): the threshold λ is calculated by minimizing the

risk function $R(k)$: $R(k) = \dfrac{N - 2k + \sum\limits_{j=1}^{k} V_2(j) + (n-k)V_2(N-k)}{N}$, where V_2 is

the vector of squared signal values organized in ascending order. The λ value is given by: $\lambda = \sqrt{V_2(k_{min})}$, where k_{min} is the index minimizing the risk functions.
- "heuristic SURE" (heur sure): a selection based on a combination of "universal" and "rigorous SURE" options.

As regards thresholding, two options can be considered. The first one is global and applies the same λ value at all the scales, while the second one is adaptive and the threshold value is computed for each scale. In this study, three threshold rescaling options are used:
- The global threshold case, which corresponds to a Gaussian white noise $N(0, \sigma^2)$, where the noise level σ is assumed to be equal to 1.
- The first scale dependent case, which is based on a single estimation of noise level σ calculated from the first-level coefficients (D1);
- The multiple scale dependent case (MSD), which consists of estimating noise level σ for each level of the wavelet transform.

In all these options, the noise level σ is considered to rescale the threshold value λ.

3.3. Application on Synthetic Data

To assess the denoising performance of the approaches suggested above (EMD denoising and DWT thresholding), simulated data were generated. The synthetic signals were modeled for different waveforms: "blocks," "heavy sine," "Doppler," and "mishmash."

The denoising performance can be quantified by computing the mean square error (MSE):

$$MSE = \frac{1}{N}\sum_{i=0}^{N-1}\left(\tilde{x}_i - x_i\right)^2.$$

The lesser the MSE value, the closer the estimated signal $\tilde{x}$ to x, and the more efficient the denoising process. Regarding wavelet thresholding, the optimal denoising parameters (thresholding function, threshold selection rule, and threshold rescaling option [see Table 3.1]) related to blocks, heavy sine,

Table 3.1 Wavelet thresholding parameters.

	Thresholding function	Threshold selection rule	Threshold rescaling option	Wavelet/level
Blocks	Hard	Universal	MSD	Db1/8
Heavy sine	Soft	Heur sure	MSD	Db3/5
Doppler	Soft	Rig sure	MSD	Sym4/5
Mishmash	Soft	Rig sure	MSD	Sym4/5

Table 3.2 MSE for EMD denoising and wavelet threshholding methods obtained from different input signals.

Signal	Method	Signal-to-Noise Ratio						
		15	10	5	3	2.5	2	1.5
Blocks	DWT	2.358	2.372	2.516	3.046	3.314	3.580	3.829
	EMD	**0.237**	**0.228**	**0.292**	**0.474**	**0.607**	**0.835**	**1.330**
Heavy sine	DWT	0.333	0.338	0.354	0.368	0.373	0.384	0.398
	EMD	**0.017**	**0.020**	**0.042**	**0.090**	**0.122**	**0.169**	**0.275**
Doppler	DWT	**0.000**	**0.000**	**0.001**	**0.002**	**0.002**	**0.003**	**0.005**
	EMD	0.017	0.014	0.010	0.008	0.008	0.008	0.010
Mishmash	DWT	0.754	0.759	0.799	0.853	0.886	0.933	1.001
	EMD	**0.753**	**0.758**	**0.786**	**0.824**	**0.851**	**0.913**	**1.000**

Note: Bolded numbers indicate the least mean square error values.

and Doppler signals are drawn from previous research [*Honório et al.*, 2012], whereas those corresponding to mishmash signals were determined in this paper after tests.

For each analyzed signal, a hundred values (100) of Gaussian white noise were carried out. For each realization, amounts of random noise are added to the noise-free signal to give SNRs of different values. Here, the SNR value is defined as the ratio of the maximum amplitude of the event to the maximum amplitude of random noise. For a given SNR value, each of the resulting noisy signals is filtered using the EMD denoising and DWT thresholding methods. The considered MSE values are calculated by averaging the values resulting from the 100 iterations (Table 3.2).

As can be seen from Table 3.2, the EMD denoising method yields the best results for blocks, heavy sine, and mishmash signals, considering the different SNR values. However, for Doppler signals, the DWT thresholding is found to be more efficient than the EMD denoising technique, and the difference between the MSE values obtained using the two studied methods is small and decreases as the SNR value gets smaller.

To illustrate the above results, we present an application of both denoising methods on heavy sine signals for different SNR values (Figure 3.1). As can be seen, the EMD denoising method outperforms the DWT thresholding in terms of efficiency.

3.4. Application on Real Seismic Data

This section is devoted to showing the results obtained from an application of the suggested denoising method on a real dataset. The data used represent real seismograms recorded during a refraction seismic survey conducted in the Algerian Sahara (Figure 3.2). The seismograms were recorded by a down-hole array of eight sensors located at depths ranging from 2.5 to 20 m with a 2.5 m separation between them. The sampling rate of the sensors was 4 ms, and the distance between the source used and the hole is 3 m.

Here, we compare the results obtained from the available real dataset using the DWT and the EMD denoising methods (Figures 3.3 and 3.4). For the former, we considered the wavelet thresholding parameters related to the four synthetic test signals analyzed in the above section. As can be noted from Figure 3.3, the obtained waveform of the seismograms denoised using the DWT depends on the chosen wavelet thresholding parameters (thresholding function, threshold selection rule, and threshold rescaling option). Moreover, compared to the DWT in the four studied situations, the EMD denoising seems to yield relatively better denoised seismograms. The suggested denoised scheme offers the advantage of being simpler and easier, and does not require to specify input parameters as the DWT does.

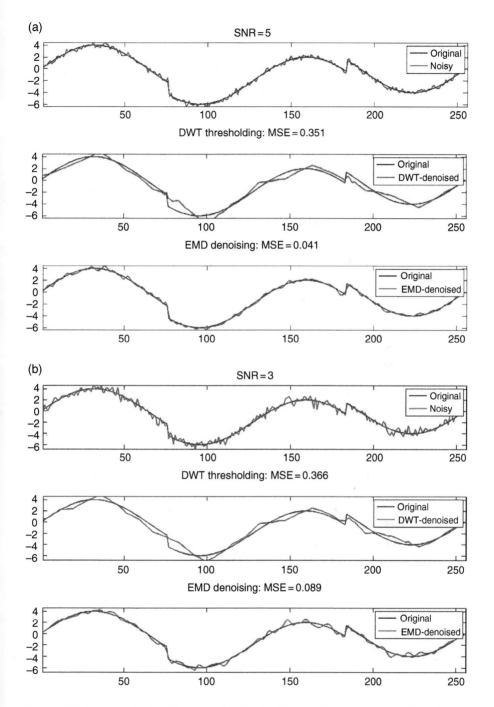

Figure 3.1 Results obtained by wavelet thresholding and the suggested EMD-based denoising technique from noisy heavy sine signal with different signal-to-noise ratio (SNR) values, SNR = 5 (a), 3 (b), and 2 (c).

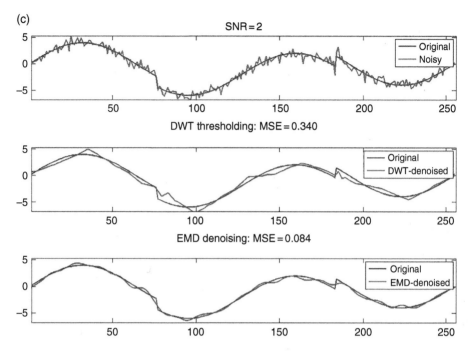

Figure 3.1 (Continued)

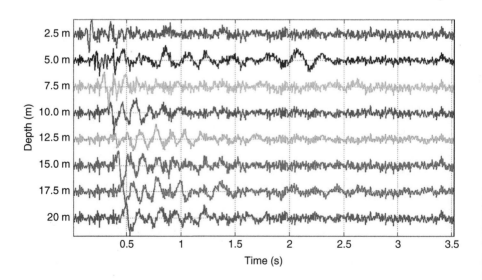

Figure 3.2 Real seismograms recorded in the Algerian Sahara.

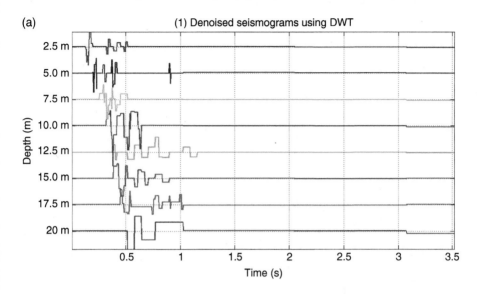

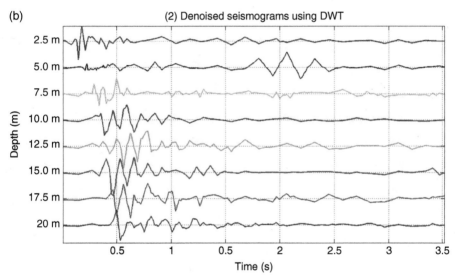

Figure 3.3 Denoised seismograms obtained by wavelet thresholding, using the parameters of Table 3.1 related to the different synthetic signals: (a) blocks, (b) heavy sine, (c) doppler, and (d) mishmash.

(c)

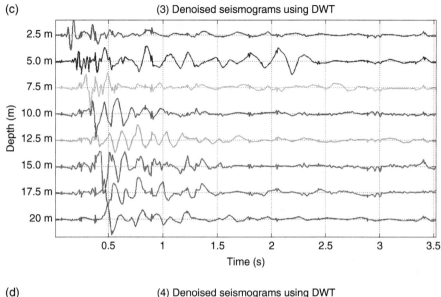

(d)

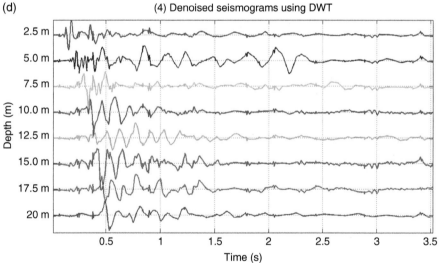

Figure 3.3 (Continued)

3.5. Conclusion

Denoising is of high importance in signal processing. In this context, a denoising technique based on empirical mode decomposition is suggested.

The results obtained from simulated data show that the suggested method outperforms the DWT thresholding for blocks, heavy sine, and mishmash signals,

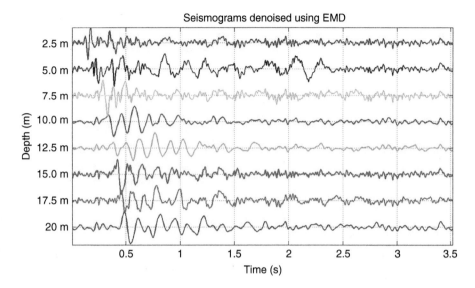

Figure 3.4 Denoised seismograms from the real dataset, using the EMD denoising method.

considering the different SNR values. However, for Doppler signals, the DWT thresholding is slightly more efficient than the EMD denoising technique, and the difference between the MSE values obtained using the two studied methods is very small.

The application on real seismograms shows that the waveforms of the signal denoised using the DWT depend on the selected wavelet thresholding parameters and are of poorer quality than those derived using EMD. In addition, a note-worthy advantage of the latter is that it does not require any preselected param-eter. To conclude, the EMD denoising technique can be successfully applied to enhance the efficiency of seismic signal denoising.

References

Daubechies, I. (1988), Orthonormal bases of compactly supported wavelets. *Commun. Pure Appl. Math.*, 41(7), 909–996.

Donoho, D., and I. Johnstone (1994), Ideal spatial adaptation by wavelet shrinkage. *Biometrics*, 81(3), 425–455.

Flandrin, P., and P. Gonçalvès (2004), Empirical mode decompositions as data driven wavelet-like expansions. *Int. J. Wavelets, Multires. Info. Proc.*, 2, 477–496.

Flandrin, P., P. Gonçalvès, and G. Rilling (2004), Detrending and denoising with the empirical mode decompositions. *Proc. European Signal Processing Conference*, pp. 1582–1584.

Flandrin, P., G. Rilling, and P. Gonçalvès (2005), EMD equivalent filter banks, from inter-pretation to applications. In *Hilbert-Huang Transform and Its Applications*, 1st edition, edited by N. E. Huang and S. Shen. World Scientific, Singapore.

Gaci, S. (2013), The use of wavelet-based denoising techniques to enhance the first arrivals picking on seismic traces, IEEE Transactions on Geoscience and Remote Sensing (TGRS), 52(8), 4558–4563, doi:10.1109/TGRS.2013.2282422.

Goupillaud, P., A. Grossmann, and J. Morlet (1984), Cycle-octaves and related transforms in seismic signal analysis. *Geoexploration*, 23(1), 85–102.

Honório, B.C.Z., R. D. Drummond, et al. (2012), Well log denoising and geological enhancement based on discrete wavelet transform and hybrid thresholding, *Energy Explor. Exploit.*, 30(3), 417–433.

Huang, N. (2005), Introduction to the Hilbert-Huang transform and its related mathematical problems. In *Hilbert-Huang Transform and Its Applications*, edited by N. E. Huang and S. Shen, pp. 1–26. World Scientific, Singapore.

Huang, N., Z. Shen, and S. Long (1999), A new view of nonlinear water waves: The Hilbert spectrum. *Ann. Rev. Fluid Mech.*, 31, 417–457.

Huang, N., Z. Shen, S. Long, et al. (1998), The empirical mode decomposition method and the Hilbert spectrum for non-stationary time series analysis. *Proc. Roy. Soc. London*, 454A, 903–995.

Huang, N., M. Wu, S. Long, et al. (2003), A confidence limit for the empirical mode decomposition and Hilbert spectral analysis. *Proc. Roy. Soc. London*, 459A(2037), 2317–2345.

Khaldi, K., M. Turki-Hadj Alouane, and A. Boudraa (2008), A new EMD denoising approach dedicated to voiced speech signals. *Signals, Circuits and Systems, SCS 2008, 2nd International Conference*, 7–9, pp. 1–5.

Mallat, S. (1989), A theory for multiresolution signal decomposition: The wavelet repre-sentation. *IEEE Trans. Pattern Anal. Mach. Intel.*, 11(7), 674–693.

Rilling, G., and P. Flandrin (2008), One or two frequencies? The empirical mode decom-position answers. *IEEE Trans. Signal Process.*, 56(1), 85–95.

Rilling, G., P. Flandrin, and P. Gonçalvès (2003), On empirical mode decomposition and its algorithms. *IEEE-EURASIP Workshop on Nonlinear Signal and Image Processing*.

Stein, C. (1981), Estimation of the mean of a multivariant normal distribution. *Ann. Stat.*, 9(6), 1135–1151.

Tsolis, G., and T. Xenos (2009), Seismo-ionospheric coupling correlation analysis of earthquakes in Greece, using Empirical Mode Decomposition. *Nonlinear Processes of Geophysics*, 16, 123–130.

Weng, B., M. Blanco-Velasco, and K. Burner (2006), ECG denoising, based on the empirical mode decomposition. *Proceedings of the 28th IEEE EMBS Annual International Conference New York City, USA*, Aug 30–Sept 3, 2006.

Wu, Z., and N. Huang (2004), Study of the characteristics of white noise using the empirical mode decomposition method. *Proc. Roy. Soc. London*, 460, 1597–1611.

4

4. A LITHOLOGICAL SEGMENTATION TECHNIQUE FROM WELL LOGS USING THE HILBERT-HUANG TRANSFORM

Said Gaci

Abstract

The scale-invariance is a characteristic of many geophysical phenomena typically containing structures with sizes that extend over several orders of magnitude in length and/or time scale and that display a certain similarity. Borehole measurements exhibit features at different scales that cannot be captured by the conventional methods. Here, we propose a new method based on Hilbert-Huang transform (HHT), a combination of the empirical mode decomposition (EMD) and Hilbert transform, for estimating a local scaling coefficient from well logs. This parameter measures the degree of heterogeneity of the layers crossed by the borehole. The proposed technique has been tested on synthetic well logs data, then applied on P- and S-wave seismic velocity logs recorded at the KTB main borehole drilled for the German Continental Deep Drilling program. The calculated depth-dependent scaling parameter highlighted the lithological heterogeneities that occurred within the logged depth interval and allowed measurement of their complexity. To conclude, the proposed technique presents a new approach to investigate multiscale features of the logs data and may enrich the conventional analysis. More datasets are necessary to draw a possible relationship between the local scaling parameter and lithology.

Sonatrach- IAP, Boumerdès, Algeria

Oil and Gas Exploration: Methods and Application, Monograph Number 72,
First Edition. Edited by Said Gaci and Olga Hachay.
© 2017 American Geophysical Union. Published 2017 by John Wiley & Sons, Inc.

4.1. Introduction

Well logs provide important information in oil and gas engineering. Their nonlinear and nonstationary properties make conventional signal processing methods unsuitable for their analysis. Hence, general data-driven techniques without a priori assumptions basis are needed to investigate such data. In this view, the empirical mode decomposition (EMD) has been proposed to explore borehole measurements.

Initially proposed by *Huang et al.* [1998], this adaptive time-frequency technique has been applied in several research fields [*Loh et al.*, 2001; *Hwang et al.*, 2003; *Huang and Attoh-Okine*, 2005; *Huang and Shen*, 2005; *Battista et al.*, 2007; *Senroy et al.*, 2007; *Yan and Gao*, 2007; *Wu and Huang*, 2009; *Gaci*, 2014a, 2014b; *Gaci and Zaourar*, 2014]. The EMD consists of decomposing the log $X(z)$ into finite basis functions called modes or intrinsic mode functions (IMFs); each IMF has a mean wavenumber. The suggested method in this study is based on the analysis of heterogeneities from well logs by investigating the exponential relation between the IMF's index and the mean wavenumber, and estimating a local scaling parameter $(\rho(z))$. The latter describes the multiscales properties of the signal and quantifies the degree of complexity of the explored geological medium.

The remainder of this chapter is structured as follows. A mathematical background of the suggested algorithm is briefly given. Then, the analysis and discussion of the results obtained from applications on synthetic and real velocity logs are presented in sections 3 and 4, respectively. Finally, the main findings and the perspectives of our research are given in the conclusion.

4.2. Theory

This section is devoted to present the mathematical background of the new Hilbert-Huang transform (HHT)-based method suggested for investigating the well heterogeneities. For recall, the HHT method combines the EMD and Hilbert transform. The EMD decomposes any nonlinear and nonstationary data set into a finite and often small number of IMFs without requiring a priori basis as in the case of Fourier and wavelet-based methods. [*Huang et al.*, 1998; *Flandrin and Gonçalvès*, 2004].

Each resulting IMF must satisfy two conditions:
1. The number of local extrema and the number of zero-crossings must be equal or differ at most by one.
2. At any time, the mean value of two envelopes determined by the local maxima and local minima is zero.

The EMD procedure, an iterative or "sifting" algorithm, can be described as follows:
1. Determine zero-crossing, local extrema maxima and minima of the signal $X(z)$.
2. Construct the upper $U(z)$ and lower $L(z)$ envelope of the signal linking, respectively, the local maxima and minima by cubic spline interpolation.

3.   Compute the local average envelope $m(z)$ of the signal: $m(z) = \dfrac{U(z) + L(z)}{2}$.
4.   Extract the detail: $h_1(z) = X(z) - m(z)$.
5.   Replace the signal $X(z)$ with $h_1(z)$, and repeat the above procedures (steps 1–4) until an IMF is obtained (the two conditions are satisfied).

The sifting algorithm is stopped by one of the these criteria: after extracting n IMFs, the residue $r_n(z)$ is either an IMF or a monotonic function. More details on stopping criteria are given in many references [*Huang et al.*, 1998; *Flandrin and Gonçalvès*, 2004; *Huang et al.*, 2008].

Finally, the EMD algorithm leads to the construction of the original signal $X(z)$ by a linear superposition:

$$X(z) = \sum_{m=1}^{n-1} \mathrm{IMF}_m(z) + r_n(z), \tag{4.1}$$

where $n-1$ is the total number of the IMF components.

After performing the EMD algorithm, the next step of HHT consists of applying Hilbert transform to each of the obtained IMF components:

$$H[\mathrm{IMF}_m(z)] = \frac{1}{\pi} PV \int_{-\infty}^{\infty} \frac{\mathrm{IMF}_m(\tau)}{z - \tau}\, d\tau, \tag{4.2}$$

where PV refers to the Cauchy principal value of this integral.

The resulting set of analytic signals can be expressed as

$$y_m(z) = \mathrm{IMF}_m(z) + j H[\mathrm{IMF}_m(z)] = a_m(z).e^{j\theta_m(z)}, \tag{4.3}$$

where $j = \sqrt{-1}$, $a_m(z)$, and $\theta_m(z)$ are respectively the instantaneous amplitude and phase function of the mth IMF:

$$a_m(z) = \sqrt{(\mathrm{IMF}_m(z))^2 + (H[\mathrm{IMF}_m(z)])^2} \tag{4.4}$$

$$\theta_m(z) = \tan^{-1}\left(\frac{H[\mathrm{IMF}_m(z)]}{\mathrm{IMF}_m(z)}\right). \tag{4.5}$$

For each IMF, differentiating the phase allows extraction of the instantaneous wavenumber:

$$\omega_m(z) = \frac{1}{2\pi} \frac{d\theta_m(z)}{dz}. \tag{4.6}$$

The original signal can be represented as a linear combination of real parts of $y_m(z)$ and a residue term $r_n(z)$

$$X(z) = \mathrm{Re}\left\{\sum_{m=1}^{n-1} a_m(z).e^{j\theta_m(z)}\right\} + r_n(z) = \mathrm{Re}\left\{\sum_{m=1}^{n-1} a_m(z).e^{j\int \omega_m(z)dz}\right\} + r_n(z), \quad (4.7)$$

where Re means real part.

Both instantaneous amplitude and wavenumber depend on the spatial position z. That allows to represent a (joint) wavenumber-space distribution of amplitude (or energy, the square of amplitude), designated as Hilbert spectrum $H(\omega, z)$.

The IMFs are considered the basis vectors representing the data. Each basis is characterized by its specific mean wavenumber (k_m), which is inversely proportional to the characteristic scale. The mean wavenumber can be estimated by considering the (energy-weighted) mean wavenumber in the Fourier power spectrum [*Huang et al.*, 1998]:

$$k_m = \frac{\displaystyle\int_0^\infty k S_m(k)dk}{\displaystyle\int_0^\infty S_m(k)dk} \quad (4.8)$$

where $S_m(k)$ is the Fourier spectrum of mth IMF mode (IMFm). Besides, *Xie and Wang* [2006] suggested another HHT-based estimation of k_m value:

$$k_m = \frac{\displaystyle\sum_{z=0}^{Z} \omega_m(z).a_m^2(z)}{\displaystyle\sum_{z=0}^{Z} a_m^2(z)}, \quad (4.9)$$

where Z is the total depth.

For a given IMFm, the mean wavenumber k_m is related to the mode number m by

$$k_m = k_0 \rho^{-m}, \quad (4.10)$$

where k_0 is a constant and ρ is computed from the least-squares fitting of the $\log(k_m)$–m plot. This relation shows that EMD behaves as a dyadic filter bank in the wavenumber domain as revealed by application on stochastic simulations of fractional Gaussian noise (fGn) [*Flandrin and Gonçalvès*, 2004; *Flandrin et al.*, 2004] and white noise [*Wu and Huang*, 2004]. For such stochastic noise data, the expected ρ value is very close to 2. It is associated with the number of scales involved in the decomposition of the signal via EMD. A small ρ value corresponds to a high number of these characteristic scales, and the converse is true also. Therefore, ρ value can be considered as a global measure of the complexity of the signal.

In order to locally investigate the multiscales features of the signal, we propose a new way to estimate the ρ value at each position $z(\rho(z))$:

$$k_m(z) = k_0 \rho(z)^{-m}, \quad (4.11)$$

where the weighted mean wavenumber $k_m(z)$ is computed using the relation (4.9) within a moving L-length-window centered in z:

$$k_m(z) = \frac{\sum_{\tau=z-L/2}^{z+L/2} \omega_m(\tau).a_m^2(\tau)}{\sum_{\tau=z-L/2}^{z+L/2} a_m^2(\tau)}. \tag{4.12}$$

In the following, in order to get a smoothed scaling log $(\rho(z))$ easy to interpret, the window length L is set, after tests, to 20 times the sampling rate.

4.3. Application on Simulated Well Log Data

Several studies have demonstrated that borehole wire-line logs may be described by nonstationary fractional Brownian motions (fBms) [*Todoeschuck et al.*, 1990; *Pilkington and Todoeschuck*, 1991]. The Hurst parameter H provides an indication about the self-similarity degree and long range dependence of the well log. From a regularity point of view, the pointwise Hölder exponent (regularity degree) is almost surely equal to H for all the samples of the fBm; the higher the H, the more regular the fBm and the more homogeneous the geological medium.

In fact, fBms are everywhere singular with the same Hölder exponent and thus are not suited to studying well logs whose local regularity varies with depth. To get rid of this limitation, logs are described rather by more generalized fractal models, specifically multifractional Brownian motion (mBm) [*Gaci et al.*, 2010].

In this application, the velocity logs $X(z)$ are modeled as mBm processes using the approach proposed by *Peltier and Lévy-Véhel* [1994]. For recall, an mBm path, with N samples, is simulated by generating N fractional Brownian motions (fBms) with the Hurst parameters H(i), i=1, ..., N, and constructed by setting

$$X(i) = \text{fBm}_{H(i)}(i), \, i=1, ..., N. \tag{4.13}$$

Here, the fBms used to simulate mBms are created by the successive random additions algorithm [*Turcotte*, 1997; *Gaci et al.*, 2010]. The depth sampling interval is set to 0.1524 m as that used in the following for the KTB sonic measurements.

The efficiency of the suggested method is checked on a synthetic log corresponding to a four-layer geological model defined by the number of samples $N=8192$, and the Hurst parameter function:

$$H(i)=0.3(\text{for } i=1,...,N/4), 0.8(\text{for } i=N/4+1,..., N/2), 0.1$$
$$(\text{for } i=N/2+1,..., 3N/4), \text{ and } 0.6 (\text{for } i=3 N/4+1,..., N).$$

From Figure 4.1, it is pointed out that the scaling log $(\rho(z))$ allows identification of the four geological layers composing the modeled log. The limit

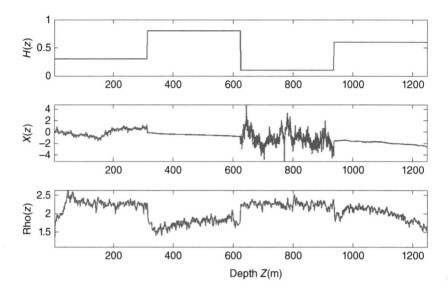

Figure 4.1 Results obtained from synthetic well log data. Top: $H(z)$: the used Hurst function. Middle: $X(z)$: the synthetic well log corresponding to a four-layer geological model and modeled as an mBm with the number of samples $N = 8192$; Hurst parameter function: H(1st layer) = 0.3, H(2nd layer) = 0.8, H(3rd layer) = 0.1, H(4th layer) = 0.6. Bottom: $\rho(z)$: the local changes of scaling ρ value.

between two adjacent layers corresponds to a ρ-value jump. In addition, it is noted that high H values correspond to small ρ values. That means more scales are needed in the EMD decomposition, and the analyzed log presents a more complex behavior. Hence, the scaling parameter ρ measures the multiscale properties of the signal and complexity of the investigated geological medium.

4.4. Application on KTB Velocity Logs

This section is devoted to the application of the suggested technique on velocity logs measured at two scientific deep boreholes: the pilot borehole (VB, 4 km), and the main hole (HB, 9.1 km), drilled in the frame of the German Continental Deep Drilling Program (KTB) (Oberpfalz, Germany). The main lithological units encountered by the wells in the studied depth intervals are amphibolite units (AU1, AU2, AU3), gneiss units (GU1, GU2, GU3), and variegated units (VU1, VU2) (Figure 4.2).

The data analyzed in this study are (a) the P- and S-wave velocity (in short, Vp and Vs, respectively) log data measured at VB borehole in the depth interval (28.194–3990.137 m), and (b) Vp and Vs log data recorded at HB borehole in the

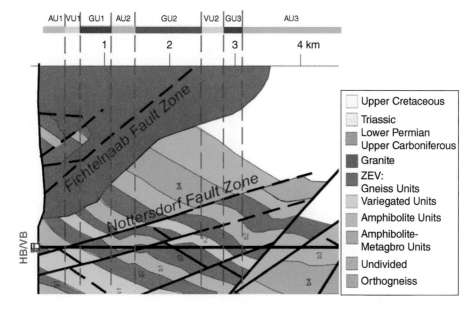

Figure 4.2 A lithologic sketch modified after *Gaci and Zaourar* [2014]. The main lithological units crossed by the wells in the analyzed depth intervals are amphibolite units (AU1, AU2, AU3), gneiss units (GU1, GU2, GU3), and variegated units (VU1, VU2).

depth interval (290.017–4509.97 m) (Figure 4.3). The sampling interval is 0.1524 m for all of the available logs.

As explained in section 4.2, prior to implementing the suggested algorithm, the HHT is applied to all the velocity well logs. The first step consists of decomposing each velocity log using the EMD, into a set of IMFs and a residue. Then the HHT is applied on the obtained IMFs corresponding to each velocity log. For illustration, Figure 4.4 presents the Hilbert amplitude spectrum corresponding to VB-Vp(z).

At an arbitrary depth $z = 2000$ m on VB-Vp log, the weighted mean wavenumber $k_m(z)$ corresponding to the mth IMF is computed using the relation (4.12). As reported in Figure 4.5, the representation of $k_m(z)$ versus the mode number m in a log-linear plan shows a power law expressed by: $k_m(z) = k_0 \rho(z)^{-m}$, where $\rho(z) = 1.960 \pm 0.012$ represents the slope of the graph. That means that EMD behaves as an almost dyadic filter bank in the wavenumber domain.

As mentioned above, the study of the variation of scaling value ρ with depth allows exploration of multiscale properties of the velocity logs, thus characterizing subsurface heterogeneities. Indeed, the ρ value is inversely proportional to the number of scales needed to decompose the signal via the EMD. The smaller the ρ value, the more complex the signal.

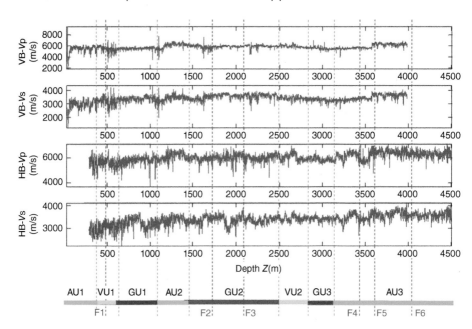

Figure 4.3 $Vp(z)$ and $Vs(z)$ velocity logs recorded at the pilot (VB) and main (HB) boreholes. The main lithological units and faults are marked on the studied depth interval.

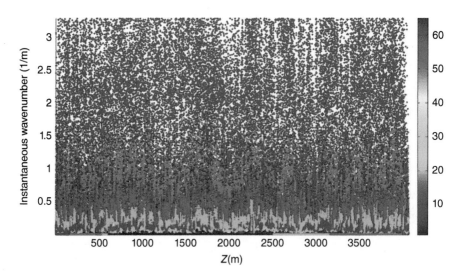

Figure 4.4 Hilbert amplitude spectrum of the velocity log $Vp(z)$ recorded at the pilot KTB borehole. The amplitude is given in m/s.

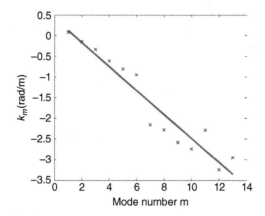

Figure 4.5 Representation of the mean wavenumber $k_m(z = 2000$ m$)$ (calculated from the Hilbert amplitude spectrum in Figure 4.4) vs. mode number m in a log-linear plan. The graph reveals a power law expressed by $k_m(z) = k_0 \rho(z)^{-m}$ with $\rho(z) = 1.960 \pm 0.012$.

Figure 4.6 illustrates that the complexity derived from the investigated velocity logs evolves with depth. For a given borehole, both local ρ logs computed from Vp and Vs logs are significantly dependent; their correlation coefficients are 0.61 and 0.74 for VB and HB boreholes, respectively.

It is noteworthy that the lithological discontinuities, either interfaces between identified geological units or fault contacts, are marked on the scaling logs by sudden changes in the estimated ρ value (red arrows point out the depths where the local extrema of the ρ logs match these lithological changes). Other in-between jumps of the ρ value are also noticed, which may be explained by local lithological discontinuities within these main units. The suggested method presents a new way to characterize subsurface heterogeneities and to identify lithological discontinuities from the scaling logs $(\rho(z))$ on which a discontinuity is recognized by a ρ-value jump.

4.5. Conclusion

This chapter proposes a new approach based on the Hilbert-Huang transform (HHT) to explore susbsurface heterogeneities from well logs using the local scaling parameter $(\rho(z))$.

This method was first tested on synthetic velocity log data, then applied on real velocity logs recorded at the main and the pilot KTB boreholes. Based on the scaling logs obtained from the analyzed velocity logs, a lithological segmentation is performed. Nearly all the lithological discontinuities are recognized on the computed scaling logs by abrupt variations of the estimated ρ value.

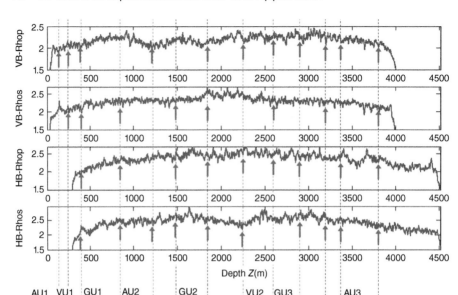

Figure 4.6 Results of the lithological segmentation obtained from the scaling logs derived from velocity logs recorded at the KTB borehole. From top to bottom: local changes of ρ value estimated from velocity logs (VB-ρP(z), VB-ρS(z), HB-ρP(z), and HB-ρS(z)), and a geological section crossing the KTB borehole (the lithologic sketch modified from http://www.gfz-potsdam.de/fileadmin/gfz/sec64/image/KTB/Figure_2.pdf). Blue line: fault contacts; grey line: lithological changes. Red arrows show the depths where the local extrema of the ρ logs match the lithological discontuities (geological unit boundaries and fault contacts).

Our findings reveal that the proposed technique is of a high importance for performing a lithological segmentation and detecting geological discontinuities using an estimated scaling log. Ample applications on more datasets are required to draw a meaning of the ρ parameter value in terms of lithology.

References

Battista, B. M., C. Knapp, T. McGee, and V. Goebel (2007), Application of the empirical mode decomposition and Hilbert-Huang transform to seismic reflection data. *Geophysics (Society of Exploration Geophysicists)*, 72(2), H29–H37.

Flandrin, P., and P. Gonçalvès (2004), Empirical mode decompositions as data driven wavelet-like expansions, *Int. J. Wavelets, Multires. Info. Proc.*, 2, 477–496.

Flandrin. P., G. Rilling, and P. Gonçalvès (2004), Empirical mode decomposition as a filter bank, *IEEE Sig. Proc. Lett.*, 11, 112–114.

Gaci, S. (2014a), A multi-scale analysis of Algerian oil borehole logs using the empirical mode decomposition. In *Advances in Data, Methods, Models and Their Applications in Oil/Gas Exploration*, edited by S. Gaci and O. Hachay, 255–276, Science Publishing Group.

Gaci, S. (2014b) A Hilbert-Huang Transform-Based Analysis of Heterogeneities from Borehole Logs, In *Advances in Data, Methods, Models and Their Applications in Oil/Gas Exploration*, edited by S. Gaci and O. Hachay, 277–295, Science Publishing Group.

Gaci, S., and N. Zaourar (2014), On exploring heterogeneities from well logs using the empirical mode decomposition, *Proceeding of EGU 2014, Energy Procedia*, 59, 44–50.

Gaci, S., N. Zaourar, M. Hamoudi, and M. Holschneider (2010), Local regularity analysis of strata heterogeneities from sonic logs. *Nonlin. Processes Geophys.*, 17, 455–466, available at http://www.nonlin-processes-geophys.net/17/455/2010/.

Huang, N. E., and N. O. Attoh-Okine (2005), *The Hilbert-Huang Transform in Engineering*. CRC Press.

Huang, N. E., and S. S. Shen (2005), *Hilbert-Huang transform and its applications*. World Scientific.

Huang, N. E., Z. Shen, S. R. Long, M. C. Wu, E. H. Shih, Q. Zheng, C. C. Tung, and H. H. Liu (1998). The empirical mode decomposition method and the Hilbert spectrum for non-stationary time series analysis, *Proc. Roy. Soc. London*, 454A, 903–995.

Huang, Y., F. G. Schmitt, Z. Lu, and Y. Liu (2008), An amplitude-frequency study of turbulent scaling intermittency using Hilbert spectral analysis. *Europhys. Lett.*, 84, 40010.

Hwang, P. A., Huang, N. E., and D. W. Wang (2003), A note on analyzing nonlinear and nonstationary ocean wave data, *Applied Ocean Research*, 25(4), 187–193.

Loh, C. H., T. C. Wu, and N. E. Huang (2001), Application of the empirical mode decomposition-Hilbert spectrum method to identify near-fault ground-motion characteristics and structural responses, *Bull. Seismol. Soc. Amer.*, 91(5), 1339–1357.

Peltier, R. F., and J. Lévy-Véhel (1994), A new method for estimating the parameter of fractional Brownian motion, *Technical report, INRIA RR* 2396.

Pilkington, M., and J. P. Todoeschuck (1991), Naturally smooth inversions with a priori information from well logs, *Geophysics*, 56(11), 1811–1818.

Senroy, N., S. Suryanarayanan, and P. F. Ribeiro (2007), An improved Hilbert-Huang method for analysis of time-varying waveforms in power quality. *Power Systems, IEEE Transactions on (IEEE)*, 22(4), 1843–1850.

Todoeschuck, J. P., O. G. Jensen, and S. Labonte (1990), Gaussian scaling noise model of seismic reflection sequences: Evidence from well logs, *Geophysics*, 55, 480–484.

Turcotte, D. L. (1997), *Fractals and Chaos in Geology and Geophysics*, Cambridge University Press, Cambridge.

Wu, Z., and N. E. Huang (2004), A study of the characteristics of white noise using the empirical mode decomposition method, *P. Roy. Soc. A-Math. Phy.*, 460, 1597–1611.

Wu, Z., and N. E. Huang (2009), Ensemble empirical mode decomposition: a noise-assisted data analysis method, *Advances in Adaptive Data Analysis* (World Scientific Publishing Company), 1(1), 1–41.

Xie, H., and Z. Wang (2006), Mean frequency derived via Hilbert-Huang transform with application to fatigue EMG signal analysis, *Computer Methods and Programs in Biomedicine*, 82, 114–120.

Yan, R., and R. X. Gao (2007), A tour of the Hilbert-Huang transform: An empirical tool for signal analysis, *Instrumentation & Measurement Magazine, IEEE*, 10(5), 40–45.

5

5. SEISMIC UNIX AND GNU OCTAVE FOR VSP DATA PROCESSING AND INTERPRETATION

Mohammed Farfour[1] and Wang Jung Yoon[2]

Abstract

Vertical seismic profiling (VSP) survey is a vital tool in subsurface imaging and reservoir characterization. In VSP survey, seismic waves are emitted using surface sources and recorded with the help of receivers clamped to the wall of the borehole. Geophysicists process the recorded data using a variety of processing tools. After processing, the data are then used to extract information that can help interpret surface seismic and infer reservoir properties. In this chapter, we introduce two free software packages used for VSP data processing. Seismic Unix and Octave are used to process raw VSP data. The processed data are compared with data obtained from commercial software. The comparison demonstrates that free software packages can be utilized to process VSP data and produce results with quality that is comparable to that produced using commercial software.

5.1. Introduction

VSP surveying is a vital tool in subsurface imaging and reservoir characterization. The technique has made significant advances since 1917 [*Fessenden*, 1917], when geophysicists lowered the first sound detective device down a borehole to better investigate rock velocity. The borehole seismic method has been historically thought of as the Check Shot survey. This survey consists

[1] *Earth Science Department, Sultan Qaboos University, Oman*
[2] *Geophysical Prospecting Lab, Energy & Resources Engineering Department, Chonnam National University, Gwangju, South Korea*

Oil and Gas Exploration: Methods and Application, Monograph Number 72,
First Edition. Edited by Said Gaci and Olga Hachay.
© 2017 American Geophysical Union. Published 2017 by John Wiley & Sons, Inc.

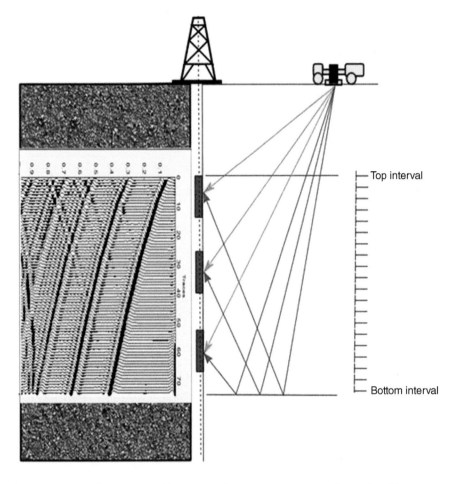

Figure 5.1 Vertical seismic profiling (VSP) downgoing (green) and upgoing (blue) waves recorded by downhole geophone. Data are recorded at predefined interval of subsurface. After recording, the geophone is moved up with a predetermined displacement distance.

of placing a receiver down the borehole at large depth intervals and sending acoustic energy down to the receiver from a surface source [*Hinds and Kuzmiski*, 2001] (Figure 5.1).

VSPs allow geophysicists to infer critical information that cannot be obtained otherwise. With VSPs, geophysicists can record waves traveling both down into the earth (direct and downgoing multiples) and back toward the surface (primaries and upgoing multiples) [*Oristaglio*, 1985]. Borehole seismic measurements have helped to overcome several difficulties faced by both seismic processers and interpreters.

They provide in situ rock properties in depth, particularly, seismic velocity, impedance, anisotropy, and attenuation. They also assist in understanding seismic wave propagation (e.g., source signatures, multiples, and mode conversions). In interpretation, VSPs are used to correct sonic logs, tie well data to seismic, and to identify key seismic markers [*Stewart*, 2001].

Depending upon the objective from the survey, there exist several VSP geometries. The most often used is the zero offset VSP, where the source is within tens of meters apart from the borehole. If the source is far away from the borehole (e.g., a few hundred meters), the survey is called offset-VSP. In case of a number of regularly offset sources from the well head, the survey is called walk-away VSP. Other VSP geometries include multiazimuth VSP, walk-around VSP, walk-above VSP, reverse VSP etc. [*Hinds et al.*, 1999].

In this chapter we focus mainly on zero-offset VSP. In zero-offset survey, a seismic energy is sent by a surface source. A geophone is clamped to the borehole wall to record different types of seismic waves. After a shot is recorded, the geophone is pulled up and clamped to an upper interval. Upon shot completion, VSP section with different wavefields is obtained [*Hinds et al.*, 1999; *Hinds and Kuzmiski*, 2001].

Open-source seismic processing packages provide a low-cost alternative to commercial software. The best-known examples of this are Stanford Exploration Project (SEP) software and Seismic Unix (SU). The latter is a free reflection processing system developed at the Colorado School of Mines. Seismic Unix has been widely used in research and teaching seismology and also in smaller-scale seismic processing in the industry. SU was first written by Einar Kjartansson in the late 1970s and released to the public in 1992 by John Stockwell at the Colorado School of Mines [*Stockwell*, 2011].

In SU commands can be written and executed in a form of script in bash language. The script is interpreted by the kernel and executed to produce results. The user can invoke several utilities to improve the execution of the desired tasks. In the Windows environment, efforts were made to develop and exploit several programming languages with graphics capabilities, and visualization to develop free toolboxes for seismic data processing, modeling, and interpretation (e.g., Matlab, Octave, SciLAB, and Python). Octave was first conceived by James B. Rawlings of the University of Wisconsin–Madison and John G. Ekerdt of the University of Texas around 1988. The program was originally intended for a chemical reactor design course. Real development was started by John Eaton in 1992. Octave is now widely used in a variety of education and research fields [*Eaton et al.*, 2015].

In this chapter we attempt to use both SU and Octave for borehole seismic data processing and interpretation. Results from processing are discussed and compared with results obtained from commercial software.

5.2. Linux: Free and Open Operating System

Unix is a hierarchical operating system that runs a program called the kernel that is the heart of the operating system. Everything else consists of programs that are run by the kernel and that give the user access to the kernel and thus to the hardware of the machine. The program that allows the user interfaces with the computer is called the "shell" [*Stockwell*, 2011]. Linux is a Unix-like computer operating system assembled under the model of free and open-source software development and distribution. The Linux kernel was first released in 1991 by Linus Torvalds. Unlike other operating systems, Linux can be freely downloaded, distributed, and modified. There are priced versions, but they are mostly cheaper than Windows. Over the years, numerous versions of Linux have been developed and received acceptance among the large scientific community (e.g., Ubuntu, Fedora, Red Hat, Debian, Archlinux, and Android). In most Linux versions, BASH (Bourne Again SHell) is the default shell. Bash is a command processor that typically runs in a text window called terminal. The user can type commands that can be interpreted as actions. It can support multiple command interpreters and also read commands from files.

5.3. Seismic Unix: Free Software for Seismic Data Processing

As SU works on a Linux environment, installing a second operating system posed a concern for many SU users. However, with the advent of virtual systems technologies, alternative solutions have been introduced in order to avoid this task. For example, Cygwin is a program that provides the user with the possibility to use a large collection of GNU and Open Source tools with functionality similar to Linux on a Windows environment. Another alternative is to install Oracle VirtualBox program on the host operating system (e.g., Windows Vista, Windows 7). It is a powerful cross-platform visualization application developed by Sun Microsystems and distributed under the GNU General Public License (GPL) or a proprietary license with additional features. VirtualBox extends the capability of the computer and host operating system so that one can run multiple guest systems at the same time. Note that the user can install and run more than one virtual machine. Under every virtual machine one can run more than one operating system. Below, detailed instructions are provided to install VirtualBox and Linux (see Table 5.1). The latest version of Ubuntu (for example) is installed as a guest system. Then SU is installed.

1. Download oracle VirtualBox Machine (VM) conformable to your windows system.
2. After installing VM, download the latest version of Ubuntu.
3. Install Ubuntu under the VM.

Table 5.1 Commands that will be used throughout the next sections are described.

Command	Use
cd	Change directory
cd ..	Go back to parent directory
chmod	Change mode of file permission
Ls	list the content of the current directory
mkdir	Create directory
rmdir	Remove a directory
pwd	Present working directory
Control –c	Kill process
|	Connect tow processes
<,>	Redirect the output of processes
&	Run process in the background

4. In Ubuntu, you create a directory as a root and change permission properties from root to user. It is a very critical step; without permission, mode change, directories, license files, and installation files cannot be created nor changed.
5. You edit the file environment in/etc as follows:
   a. Change your working directory to/etc then open the file using the text editor gedit by typing "sudo gedit environment"
   b. Include the path to your SU folder by adding "**:/cwp/bin** " to the end of the path line. Then define CWPROOT by adding the line "**CWPROOT=/cwp**"
6. Log out and log back in again. To verify if your changes work, type: "echo $PATH"
7. Some additional packages are required before installing SU. You can download and install these packages using the command "sudo apt-get install"List of packages that are needed includes:
   a. libx11-dev, and libxt-dev for the X-toolkit applications
   b. gfortran for Fortran compilation
   c. freeglut3-dev, libxmu-dev, and libxi-dev for Mesa/Open GL
   d. libc6 for libswputils (not essential)
   e. libmotif4, libxt6, x11proto-print-dev, libmotif-dev for Motif
8. Finally, you can run your installation commands one by one in the following order:
   • make install
   • make xtinstall
   • make xminstall
   • make finstall
   • make mglinstall
   • make utils

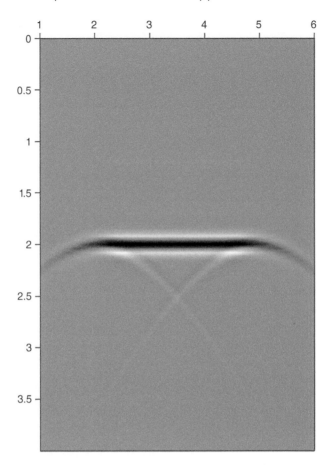

Figure 5.2 Output of susynlv.

The installation can take several minutes. To check that SU is successfully installed on the VM one can open a terminal and type any SU command. For example, you can type:

```
susynlv | suximage &
```

The first command creates a synthetic seismogram for linear velocity function. The second plots the output of the first command in an image form (Figure 5.2).

For more commands and key words the user can simply type "suname." This command lists all the programs and libraries with a short description. "sudoc" followed by the program name gives documentation of the program. Useful scripts can be found in the demos directory located in the SU installation directory [*Forel et al.*, 2005].

5.4. UNIX Stream Editor (SED) and AWK

AWK is a C-like interpreted programming language designed for text processing. It was originally developed to meet the limitations of the shell scripts in applications that call for data manipulation and reduction [*Dougherty and Robbins*, 1997]. AWK is a standard feature of most Unix-like operating systems. In SU, AWK is used to manipulate information extracted from the header (e.g., time picks, trace numbers, receiver elevation, etc.) using SU commands. The information can then be saved in a form usable in further processing, visualization, and filtering operations. For example, one can use the following command to convert time picks in column 2 from second to millisecond and save the picks into a new text file called timesec.txt:

```
awk timepicks.txt '{ print $2*1000 }'> timems.txt
```

SED is a text editor that performs editing operations on information extracted from input or file. SED is used in SU to edit, add, and remove information or text words extracted from headers [*Dougherty and Robbins*, 1997]. In the example below, "sed" is used to remove header key words from the file extracted from the header using "sugethw" and to keep only time picks and trace numbers. Picks can then be saved in a text file and used later on for further processing.

```
sugethw < ZnPick.su key=tracl,lagb | sed -e's/tracl=
//' -e's/lagb=//'
```

"awk" and "seed" can be combined to undertake some special operations. For example, the following combination is used for the same purpose as above; the time picks are then converted to second and stored in a text file.

```
sugethw < ZnPick.su key=tracl,lagb | sed -e's/tracl=//'
-e's/lagb=//'|awk '$0' | awk '{print $1,$2/1000}'>
picks.txt
awk '$0'
```

is used to delete space between lines, -e is used for multiple use of sed.

5.5. GNU Octave

Octave is an open-source interactive software system that is primarily used for numerical computations and graphics (Figure 5.3). It is particularly designed for matrix computations: solving simultaneous equations, computing eigenvectors and eigenvalues, and so on [*Eaton et al.*, 2015]. Octave syntax is developed to be similar to Matlab. After decades of improvements, Octave became an excellent alternative to Matlab. In addition, the fact that the software can be freely redistributed, copied, or modified under the terms of GNU GPL gives Octave a wide

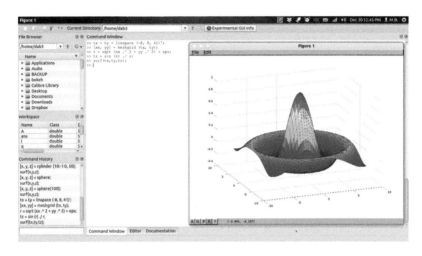

Figure 5.3 Octave interface. Octave was written by John W. Eaton et al. (2015).

acceptance among scientists and engineers. Several versions of Octave have been developed and released. The most recent version is Octave 4.0.0. This last version brings many new features and better compatibility with Matlab.

5.6. VSP Data Processing in SU and Octave

A Segy-formatted dataset was loaded to the SU for processing and interpretation. Figure 5.4 displays the processing workflow run to the data. The data were first subject to AGC to visualize different signals composing the VSP raw section. After energy balancing, the first break picking script is run to pick the time at which direct waves arrive to the downhole geophone. The user can select and zoom in the first break wave interval (e.g., 0 to 0.7 s) to see clearly the signal waveforms. After positioning the mouse cursor at the appropriate location, the 'S' key must be hit to save the pick coordinates. Upon completion, hitting the 'Q' key will close the figures. Picks are then stored in the trace header. Next, the stored picks are used to define and remove noise arriving before the first breaks (Figure 5.5). The user can also let the auto-picking script perform the picking. Note that AWK and SED play a key role in this stage. They help manipulate picks, sort them, and store them in text files and extract them whenever is necessary. The downgoing wavefields are flattened to ease their separation from upgoing waves. Figures 5.6 and 5.7 show data before and after wave separation using a median filter. By doubling the first break times, upgoing waves with remaining noise are flattened to two-way time (Figure 5.8). After several tests, 11-trace median filter is applied to the upgoing data to remove tube waves and random noise that contaminate the

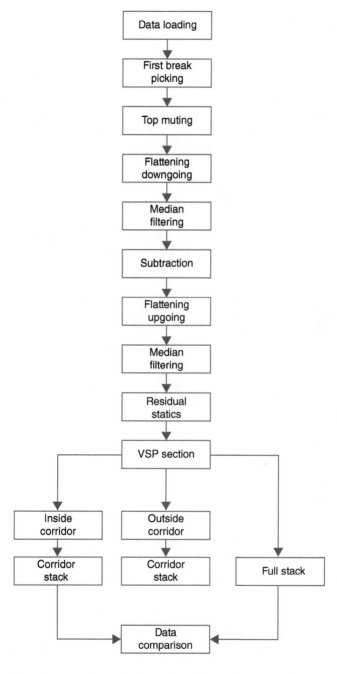

Figure 5.4 Different processing and interpretation operations carried out on the data.

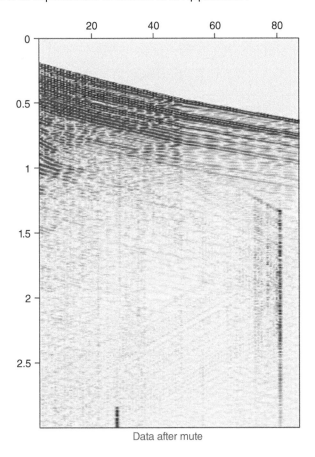

Data after mute

Figure 5.5 Raw data after scaling to improve visibility.

data. In addition, some residual high-frequency noise is attenuated using band pass frequency filtering (Figure 5.9). Note that there are more advanced filtering processes that can be carried out using SU, such as frequency-wavenumber (f-k) filtering, tau-p filtering, predictive deconvolution, etc. This richness indeed makes SU powerful and useful for a variety of seismic applications.

The same workflow is applied to the data using Octave functions and scripts. After scaling the data, first breaks are manually picked. The user selects and clicks points at desired locations. The picks are then stored and used to define the upper limit for top muting and to flatten the downgoing waves. After the wave separation using median filtering, residual static corrections are run to the data to improve its alignment and flattening. Finally, different mutes (top and bottom) are applied to the data to produce different corridors.

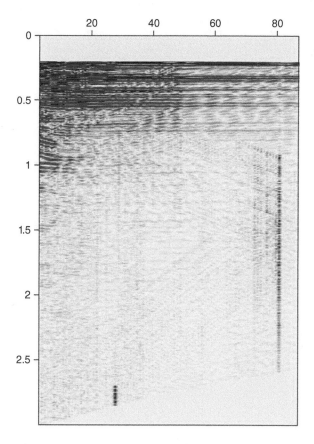

Figure 5.6 Flattened data after removing time delay.

Note that processing scripts in Octave are much simpler compared to SU. The results from processing are shown in Figures 5.10 to 5.16. All necessary operations can be carried out using Octave toolboxes. Users can find a variety of tools and built-in functions that can ease processing and improve visualization. Note that most of the functions and script that run in Matlab can be used in Octave. Thus, several toolboxes developed for digital signal processing, seismic signal processing, and image processing can be integrated easily in Octave. In fact, this helped us load several Matlab functions and tools to complement the processing workflow.

To evaluate the reliability and robustness of the processing results from workflows above, the raw data is processed using widely used commercial software. Final VSP sections from SU and Octave are compared with the data from the commercial software (Figures 5.16 and 5.17). Interestingly, datasets are found to

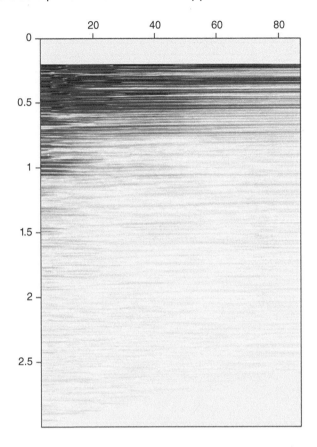

Figure 5.7 Downgoing waves after median filtering.

be very consistent to one another. However, the fact that SU and Octave are fully free of charge is indeed a great advantage over the commercial software.

5.7. VSP Data Interpretation

VSP data have several applications that can be of great interest to surface seismic processors and interpreters. The VSP data can be used in seismic processing and interpretation to estimate the phase of the surface seismic data, calculate attenuation that seismic waves undergo by going to deeper formations in subsurface, recognize multiples, and estimate different seismic velocities [*Campbell et al.*, 2005]. In the example presented here we use VSP data to detect multiples, which are seismic events that reflect more than once. Basically, primary upgoing waves intersect with first-break downgoing event at the location of the reflector

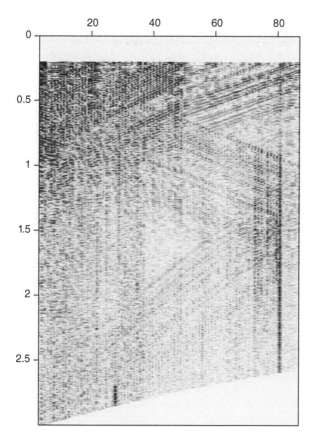

Figure 5.8 Upgoing waves along with different types of noise that contaminate the data.

that creates the upgoing events. Multiples come only after their corresponding reflections. This means that they cannot reach the primary downgoing wave (Figure 5.18). Thus, one can easily identify them on VSP data. This provides a great assistance to surface seismic where discriminating primaries from multiples poses a real challenge in both processing and interpretation. This can also be done by carefully defining inside and outside corridors. The outside corridor starts with upgoing waves and extends to a few dozen milliseconds in width so that only primaries are encompassed. The corridors are stacked and repeated several times for visualization. Comparing the corridors with the VSP section can show clearly the multiples. In Figure 5.19, comparison between the magnified outside corridor stack (middle) and the VSP section (right) shows that multiples (identified in red color) are absent in the corridor stack, whereas primaries are strongly present (yellow arrow).

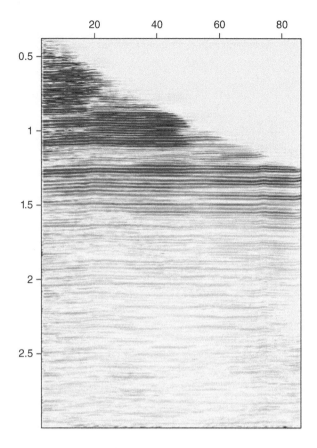

Figure 5.9 Upgoing waves after conversion to two-way time. The data are now ready for correlation with surface seismic.

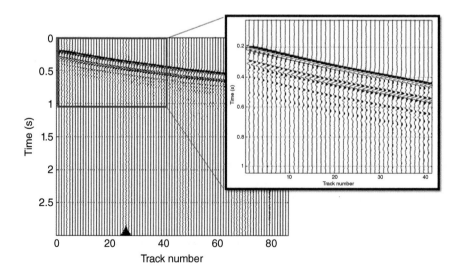

Figure 5.10 Energy balancing of traces to correct the data for geometrical spreading.

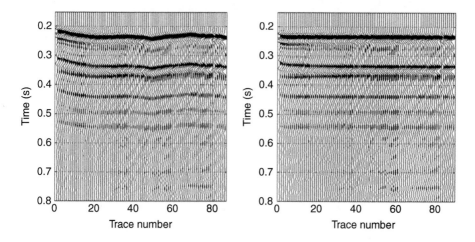

Figure 5.11 Residual statics applied to the data to improve its alignment.

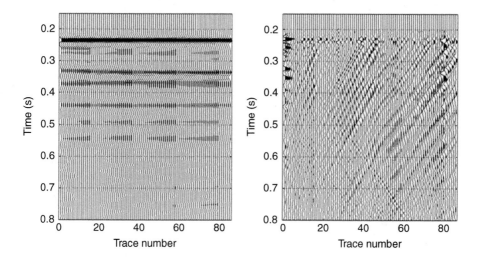

Figure 5.12 Downgoing and upgoing waves after separation using median filter.

5.8. Discussion and Conclusion

VSP data are rich in information about geological formations in subsurface. To fully exploit the data from VSP, numerous processing and filtering processes are needed. Different software packages for VSP data processing and interpretation are available on the market. There are several free and open programs that are rich in functions and tools that enable the user to address VSP data as any

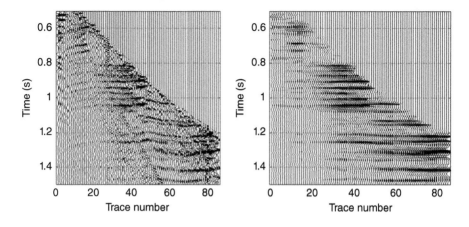

Figure 5.13 Improving the alignment of upgoing reflections and removing remaining noise.

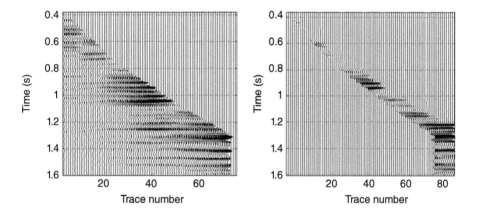

Figure 5.14 Corridor muting for interpretation purposes.

other signal. The latter programs have been successfully deployed for VSP data processing and interpretation. SU is powerful free software that is widely used in small-scale surface seismic data processing. The software can be used to process VSP data processing and provide results that are comparable to that obtained using commercial software packages. Limitations of the software lie in its installation, which cannot be easily completed except by Linux practitioners or programmers. In addition, during the processing, the user is required to utilize several complementary tools in Linux and text editors to manipulate the data files and information input and output to headers. Most of the difficulties encountered in

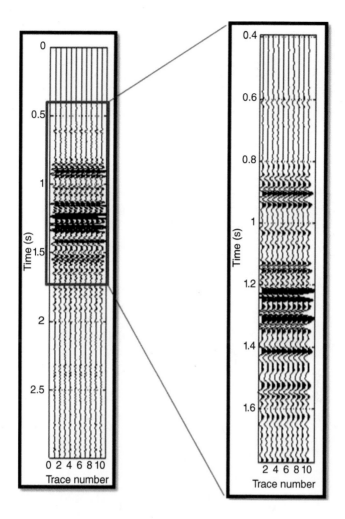

Figure 5.15 Outside corridor after stacking. The stack traces is repeated 10 times.

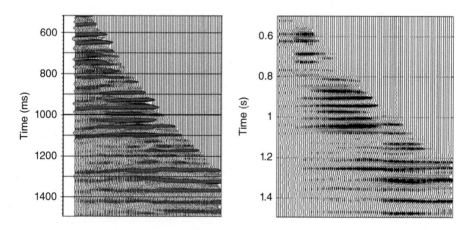

Figure 5.16 Comparison of processing results. The first (left) data is processed using commercial software, whereas the second (right) is processed using Octave GNU.

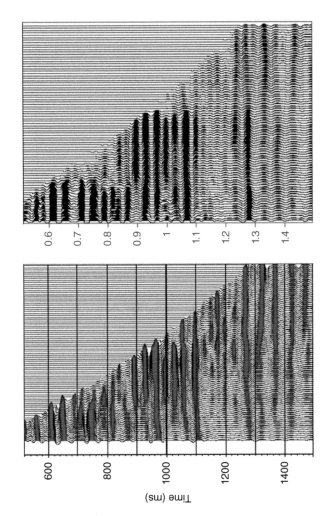

Figure 5.17 A comparison between VSP data processing results from SU (left) and commercial software (right).

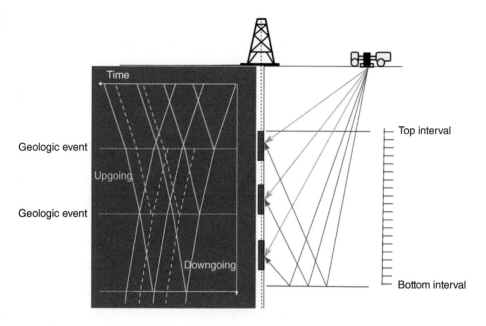

Figure 5.18 Only primary upgoing waves intersect with first break downgoing wave. Primaries (dashed blue lines) do not reach first break downgoing wave.

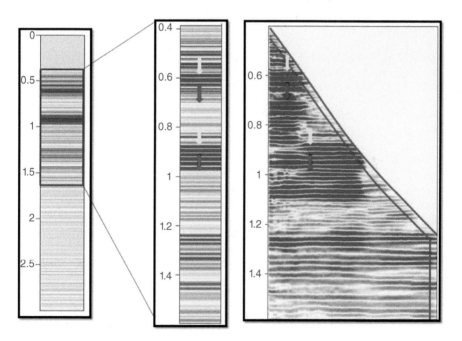

Figure 5.19 Multiple identification using VSP. Only primary waves are observed in the corridor stack (yellow arrow). The multiple events do not reach the corridor defined by the red lines; thus, multiple are not present in the corridor stack.

Table 5.2 Major differences between SU and GNU Octave.

	SU	Octave
Platform	Linux / Unix	Windows, Linux, Mac
Installation	Multi-stage process	Executable file
Classification	Free and Open Source	Free and Open Source
Functions	Non-modifiable	Modifiable
Scripts	Modifiable	Modifiable
Specification	SU is primarily intended for seismic data processing	Mainly for numerical analysis with possible extension to variety of applications and fields
Language	Written in the C language	Written in the C++ language
Latest version	SU 44 (released on 12 August 2015)	Octave 4.0.0 (released on 28 May 2015)

SU are almost overcome in GNU Ocatve (Table 5.2). The latter is also free and open software that runs under different platforms. Octave is full of built-in functions, graphical tools, and matrix and vector manipulations modules that allow the user to handle VSP data very easily compared to SU.

References

Campbell, A., A. Fryer, and S. Wakeman (2005), Vertical seismic profiles: More than just a corridor stack. *The Leading Edge* 24(7), 694–697.

Dougherty, D., and A. Robbins (1997), *sed & awk*, 2nd ed., O'Reilly Media, Beijing, p. 434.

Eaton, J., D. Bateman, S. Hauberg, and R. Wehbring (2015), *GNU Octave version 4.0.0 manual: A high-level interactive language for numerical computations*. CreateSpace Independent Publishing Platform.

Fessenden, R. A., (1917), Method and apparatus for locating ore bodies. U.S. Patent No. 1,240,328.

Forel, D., T. Benz, D. Wayne, and W. Pennington (2005), *Seismic data processing with Seismic Unix*. Society of Exploration Geophysicists, Tulsa, Oklahoma.

Hinds, R., and R. Kuzmiski (2001), VSP for the interpreter/processor for 2001 and beyond: Part 1. CSEG Recorder, September, 85–95.

Hinds, R. C., N. L. Anderson, and R. D. Kuzmiski (1999), *VSP interpretive processing: Theory and practice*. Society of Exploration Geophysicists, Tulsa, Oklahoma.

Oristaglio, M. L. (1985), A guide to current uses of vertical seismic profiles. *Geophysics*, 50, 12, 2473.

Stewart, R. (2001) VSP: An in-depth seismic understanding, *CSEG Recorder*, September, 79–83.

Stockwell, J. (2011), A course in geophysical image processing with Seismic Unix: *GPGN 461/561 Lab Fall 2011*.

6. SEISMIC SPECTRAL DECOMPOSITION APPLICATIONS IN SEISMIC: A REVIEW AND APPLICATION

Mohammed Farfour[1], Jalal Ferahtia[2], Noureddine Djarfour[3], and Mohand Amokrane Aitouch[2]

Abstract

Time-frequency analysis or spectral decomposition is a technique that allows geophysicists to visualize frequency content of seismic data along a time axis. The technology started as a simple band filtering technique that shows amplitude spectrum at user-defined bandwidth and became a routine technique in modern workflows of seismic interpretation and reservoir characterization. One advantage of this innovative approach over other processing and analysis algorithms is its simplicity and ease of use. In fact, spectral decomposition does not require advanced processing workflows; it can be programmed and run in common open software programs. Over the last decades, numerous techniques of time-frequency analysis and case studies have been published in literature. In this work after introducing the fundamentals behind the most common decomposition methods we show some real examples using Matlab software. We then use commercial software to address challenging seismic data from South Texas to reveal some features of the reservoir that are hidden in the seismic broadband.

[1] Earth Science Department, Sultan Qaboos University, Oman
[2] University of M'hamed Bougara, Boumerdès, Algeria
[3] University of Ahmed Draia, Adrar, Algeria

Oil and Gas Exploration: Methods and Application, Monograph Number 72,
First Edition. Edited by Said Gaci and Olga Hachay.
© 2017 American Geophysical Union. Published 2017 by John Wiley & Sons, Inc.

6.1. Introduction

Seismic signal can be considered a nonstationary signal composed of different components, all localized at different times and frequencies. The amplitude spectrum of Fourier transform indicates the presence of different frequencies but does not show their temporal distribution. Therefore, short time Fourier transform (STFT) was suggested. In STFT analysis, signal through a short window is considered a stationary. By shifting the time window along the whole signal, one can obtain the frequency content in a frequency-time (2-D) image [*Chakraborty and Okaya*, 1995]. Over the past decades, several analysis approaches have been developed and showed successful applications in both seismic data processing and interpretation. The majority of these algorithms belong to either linear or quadratic methods.

In the linear methods, the signal is subject to inner product with preassigned mathematical functions derived from a mother function by simple operations; examples include STFT [*Allen*, 1977], wavelet transform [*Chakraborty and Okaya*, 1995, *Sinha et al.*, 2005], and S-transform [*Stockwell et al.*, 1996]. In all these methods, the time-frequency representation can be influenced by the choice of the mathematical function with which the signal is decomposed. In addition, the Heisenberg uncertainty limits the resolution of the T-F resulting images. In quadratic methods, the problem introduced by the use of the mathematical functions is overcome. However, the interference term renders the Time-Frequency representation ambiguous and its interpretation more complicated. Furthermore, the reconstruction of the signal is more complicated than in the linear methods [*Daubechies et al.*, 2011]. Recently, other techniques have been proposed to solve the problems above. These include the matching pursuit method [*Liu and Marfurt*, 2007], local time-frequency decomposition [*Liu and Fomel*, 2013], the empirical mode decomposition families [*Han and Van der Baan*, 2013], and the synchronsqueezed transforms [*Herrera et al.*, 2014].

6.2. Time-Frequency Analysis in Reflection Seismic

For geophysicists, spectral decomposition is very similar in principle with common depth point (CDP) gathers sorting and stacking. In fact, the decomposition of a single trace produces something like a CDP gather, but instead of having offset-time representation one gets frequency-time representation. This makes the technique easy to understand and to apply. As a result, spectral decomposition techniques have received wide acceptance and popularity among exploration geophysics community. Spectral decomposition was first used for data processing in a form of time-variant frequency filtering where the full spectral bandwidth is described for each time and used to distinguish between different types of superimposed seismic events. In this section, we focus mainly on the most widely used techniques in seismic interpretation.

6.2.1. Short Time Fourier Transform–Based Analysis

The equation for the short window discrete Fourier transform can be written as

$$U_{STFT}(\tau, f) = \frac{1}{\sqrt{2\pi}} \int u(t) w(t-\tau) e^{-j2\pi ft} dt, \tag{6.1}$$

where $u(t)$ is the time domain seismic data, τ is the center time of the window function $w(t-\tau)$, f is the frequency, and $U_{STFT}(\tau,f)$ is the time-frequency function.

The defined window $w(t-\tau)$ can be either a tapered or untapered rectangular window (boxcar), Gaussian window, Hamming window, or Hanning window [*Mallat*, 2009].

Partyka et al. [1999] used a tapered rectangular window, while *Mallat* [2009] used a Gaussian window of the form:

$$w(t-\tau) = e^{-\sigma^2(t-\tau)^2}, \tag{6.2}$$

where σ is a constant value controlling the window size, with larger values of σ resulting in smaller time windows.

The spectral energy is distributed in time over the length of the window, thereby limiting resolution. If the time window is too short, the spectrum is convolved with the transfer function of the window, and frequency localization is lost (i.e., the frequency spectrum is smeared). This can be mitigated to some extent by tapering the window, but it is obviously preferable to avoid windowing altogether [*Daubechies et al.*, 2011]. Another disadvantage of a short window is that the side lobes of the arrivals appear as distinct events in the time-frequency analysis. If the time window is lengthened to improve frequency resolution, multiple events in the window will introduce notches that dominate the spectrum. Long windows thus make it very difficult to ascertain the spectral properties of individual events.

6.2.2. Wavelet Transform-Based Analysis

The continuous wavelet transform (CWT) is an example of the wavelet transform (WT) technique. WT was first introduced in the work of *Morlet et al.* [1982] and *Goupillaud et al.* [1984] but received full attention of the signal processing community when *Daubechies* [1988] and *Mallat* [1989] established connections of the WT to discrete signal processing. CWT method [*Sinha et al.*, 2005] does not require preselecting a window length and does not have a fixed time-frequency resolution over the time frequency space; it rather uses dilation and translation of a wavelet to produce a time-scale map. A single scale encompasses a frequency band and is inversely proportional to the time support of the dilated wavelet.

The CWT is defined mathematically as the inner product of the family of wavelets $\psi_{\sigma,\tau}(t)$ with the signal $u(t)$.

$$S_\omega(\sigma,\tau) = \frac{1}{\sqrt{\sigma}} \int_{-\infty}^{\infty} u(t)\bar{\psi}\left(\frac{t-\tau}{\sigma}\right)dt, \tag{6.3}$$

where $\bar{\psi}$ is the complex conjugate of ψ and S_ω is the time scale map (scalogram) used to extract the instantaneous frequency. τ is the time shift applied to the mother wavelet, and σ is scale.

In this study we use the Morlet wavelet, one of the most commonly used wavelets in seismic spectral decomposition. The Morlet mother wavelet is defined as

$$\psi_0(t) = \pi^{-1/4} e^{i2\pi t} e^{-t^2}. \tag{6.4}$$

6.2.3. S-Transform–Based Analysis

The S-transform is proposed by *Stockwell et al.* [1996] as an extension to the Morlet wavelet transform. The mother wavelet for the S-transform is also a modulated Gaussian function, but it keeps the modulation part with no scaling and no shifting. The S-transform is then defined as

$$S(\tau,f) = \int u(t) g_f(t-\tau) e^{-j2\pi ft} dt, \tag{6.5}$$

where $u(t)$ is the time domain seismic data and g_f is Gaussian function defined as

$$g_f(t) = \frac{|f|}{\sqrt{2\pi}} e^{-(ft)^2}. \tag{6.6}$$

Due to some limitations associated with invariant window, a more generalized form was proposed. In the generalized S-transform, g_f is rewritten as

$$g_f(t) = A|f| e^{-\alpha(ft-\beta)2}, \tag{6.7}$$

where A, α, and β are constants to generate a variety of window forms that can better correlate with signal.

6.2.4. Matching Pursuit Method

The Matching Pursuit Decomposition technique [*Liu and Marfurt*, 2007] also attempts to decompose a signal to constituent wavelets or atoms, selected from a wavelet dictionary, a large collection of Gabor wavelets covering the full range of time, frequency, scale, and phase index. However, in seismic, it is common practice to use the Morlet wavelet, which is considered a good approximation to a real seismic wavelet [*Wang*, 2007]. In time domain and frequency domain, the Morlet wavelet [*Morlet et al.*, 1982] is defined as

$$m_t(t) = e^{-\frac{\ln 2}{k}(ft)^2} e^{i2\pi f_m t}, \tag{6.8}$$

where f_m is the mean of the dominant frequency and k is a constant value that controls the wavelet breadth. By using different values of k, more cycles will be included in the Morlet wavelet.

The assumption behind the decomposition using matching pursuit is that a seismic trace $s(t)$ can be seen as a linear combination of Morlet wavelets and random noise.

$$s(t) = \Sigma a. m(f, d, \varphi) + \text{Noise} \tag{6.9}$$

To obtain amplitudes a and time dilations d, we exploit complex attribute analysis. The time dilation d is the time of peak envelope. The phase angle φ is calculated from instantaneous phase. The instantaneous frequency equals the mean frequency of Morlet wavelet.

6.2.5. Wigner-Ville Distribution

In contrast with the linear time-frequency representations that decompose the signal on elementary components, the purpose of the energy distributions is to distribute the *energy* of the signal over the two description variables: time and frequency [*Auger et al.*, 1996]. The Wigner distribution, almost as well known as the spectrogram, was actually introduced in the area of quantum mechanics and not at all in signal analysis [*Wigner*, 1932]. It was presented again by J. Ville in 1948 as a quadratic representation of the local time-frequency energy of a signal. The Wigner-Ville Distribution is defined as

$$W_u(t, f) = \int_{+\infty}^{-\infty} u\left(t - \frac{\tau}{2}\right) u^*\left(t - \frac{\tau}{2}\right) e^{-j2\pi f\tau} d\tau, \tag{6.10}$$

or equivalently,

$$W_u(t, f) = \int_{+\infty}^{-\infty} U\left(t - \frac{\gamma}{2}\right) U^*\left(t + \frac{\gamma}{2}\right) e^{-j2\pi\gamma t} d\gamma, \tag{6.11}$$

where U is Fourier transform of the signal u.

In analogy to the STFT, the window is basically a shifted version of the same signal. It is obtained by correlating the signal with itself at different frequencies and times. The drawback of the Wigner distribution is that it gives cross-terms that are located between and can be twice as large as the different signal components [*Auger et al.*, 1996]. A large area of research has been devoted to reduction of these cross-terms.

6.2.6. Empirical Mode Decomposition–Based Analysis

By definition, the empirical mode decomposition (EMD) is a fully data-driven technique that decomposes a given signal into a set of elemental oscillations called intrinsic mode functions (IMFs) [*Huang et al.*, 1998]. The IMFs are computed recursively, starting with the most oscillatory one. The decomposition method uses the envelopes defined by the local maxima and the local minima of the data series. Once the maxima of the original signal are identified, cubic splines are used to interpolate all the local maxima and construct the upper envelope. The same procedure is used for local minima to obtain the lower envelope. Next, one calculates the average of the upper and lower envelopes and subtracts it from the initial signal. This interpolation process is continued on the remainder. This sifting process terminates when the mean envelope is reasonably zero everywhere, and the resultant signal is designated as the first IMF. The first IMF is subtracted from the data and the difference is treated as a new signal on which the same sifting procedure is applied to obtain the next IMF. The decomposition is stopped when the last IMF has small amplitude or becomes monotonic [*Han and van der Baan*, 2013]. After that, Hilbert transform is invoked to calculate the instantaneous frequencies of the IMFs. The technique has also some obstacle restricting its performance due to mode mixing. Mode mixing is defined as a single IMF consisting of signals of widely different scales or a signal of a similar scale residing in different IMF components [*Herrera et al.*, 2014]. Thus, more advanced EMD versions were introduced such as ensemble and complete ensemble EMD [*Wu and Huang*, 2009; *Torres et al.*, 2011]. The latter methods resolve problems posed in EMD by adding Gaussian white noise onto the target signal and decomposing the signal using different noise realizations.

6.2.7. Synchron-Squeezed Transform (SST)–Based Analysis

Synchron-squeezing transform provides a way to decompose a signal $s(t)$ into constituent components with time-varying harmonic behavior. These signals are assumed to be the addition of individual time-varying harmonic components, yielding

$$s(t) = \sum A_k(t) . cos(\theta_k(t)) + ad(t). \tag{6.12}$$

Where $A_k(t)$ is the instantaneous amplitude, $ad(t)$ is the additive noise, k is the maximum number of components that form the signal, and $\theta_k(t)$ is the instantaneous phase of the components. The instantaneous frequency of the components is estimated from the instantaneous phase [*Daubechies et al.*, 2011]. The SST applications were extended recently to include seismic time-frequency analysis [*Herrera et al.*, 2014]. This new decomposition tool produced a high-resolution time-frequency representation and helped better characterize hidden properties in seismic signal than did complete ensemble EMD, because SST has the ability to adapt the mother wavelet to targeted signal.

6.3. Application Using Matlab

The use of Matlab (short for Matrix Laboratory) in research and educational applications is rapidly increasing. The program provides a rich technical computing environment that combines numerical computation, visualization, and a higher-level programming language. Using Matlab, we create first a signal that is composed of different components. The mathematical expression of the signal is illustrated below:

$$s(t) = \begin{cases} \cos(2\pi t) & for \quad 0 < t < 2.45 \\ 1 & for \quad 4.95 < t < 5.95. \\ \cos(2\pi 2t) & for \quad else \end{cases} \tag{6.13}$$

The signal is subject to time-frequency analysis using different decomposition tools in Matlab [*Auger et al.*, 1996, *Thakur et al.*, 2013] to determine its frequency components that compose the signal. Comparison of the images obtained from the process show that time and frequency resolutions differ from one algorithm to another (Figure 6.1). For example, STFT (middle) shows good lateral resolution; however, vertical resolution is so low that one cannot resolve the different frequency events in the image. On the other hand, the wavelet transform (right) shows good vertical separation, but lateral resolution is poorer than STFT. Images displayed in Figure 6.2 show that the synchron-squeezed Fourier and wavelet transforms (SSFT and SSWT) have improved significantly both the vertical (time) and lateral (frequency) resolutions. Now the frequency components of the signal are clearly seen: the shallower one is 1 Hz and the deeper is 2 Hz.

Next, we apply the different spectral decomposition techniques to real seismic data from Alberta, Canada. The data are from a hydrocarbon-bearing sand reservoir located at shallow depth. The time frequency analysis is run to traces, one by one. This creates time-frequency gathers similar to CMP (common midpoint) gathers (Figure 6.3). The images of all traces are rearranged and resorted to produce iso-frequency images for the whole section at every frequency component. The analysis of the iso-frequency images can be done in an iterative way by selecting a frequency increment so that images can be investigated at regular steps (Figures 6.4 to 6.12).

Note that the seismic response of the reservoir is readily seen on stack and prestack sections in a form of bright seismic amplitude. Several frequency

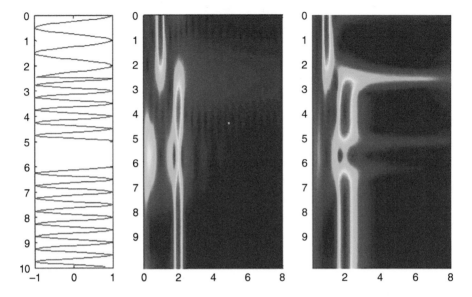

Figure 6.1 The time-frequency analysis of the synthetic signal (left). Short-time Fourier transform (STFT, middle) shows a good lateral resolution, while the wavelet transform (WT, right) shows relatively better vertical resolution.

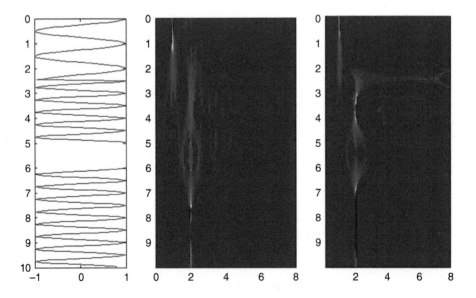

Figure 6.2 The time-frequency analysis of the synthetic signal (left). Both synchron-squeezed Fourier transform (SSFT, middle) and synchron-squeezed wavelet transform (SSWT, right) show good lateral and vertical resolution.

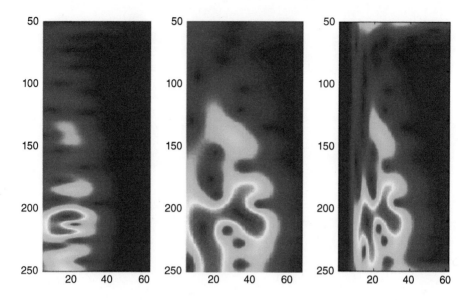

Figure 6.3 Time-frequency images of real seismic trace computed using STFT (left), WT (middle), and Stockwell transform (ST, right).

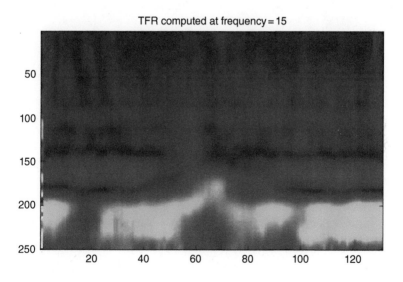

Figure 6.4 Time-frequency images computed using STFT at 15 Hz.

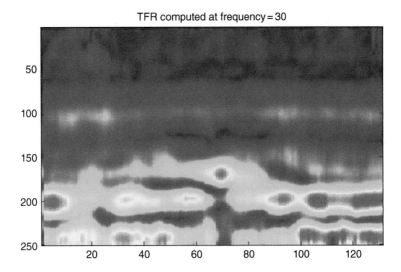

Figure 6.5 Time-frequency images computed using STFT at 30 Hz.

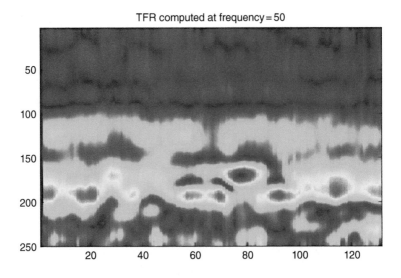

Figure 6.6 Time-frequency images computed using STFT at 50 Hz.

components are calculated using the spectral decomposition algorithms programmed in Matlab. Images obtained from the decomposition using different techniques are presented in Figures 6.4 to 6.12. The frequency images showed that the reservoir demonstrates a frequency response that is different relative to

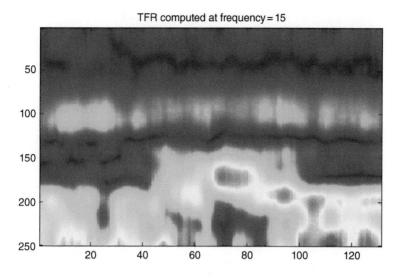

Figure 6.7 Time-frequency images computed using continuous wavelet transform (CWT) at 15 Hz.

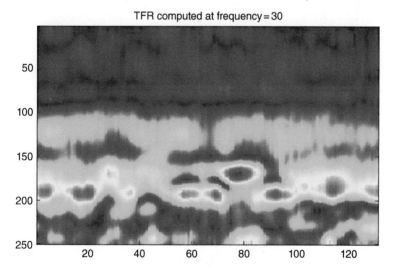

Figure 6.8 Time-frequency images computed using CWT at 30 Hz.

TFR computed at frequency = 50

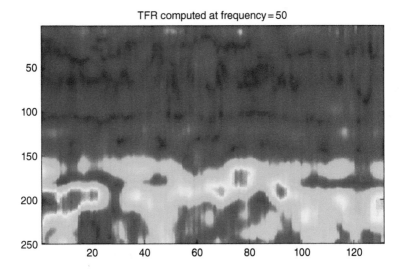

Figure 6.9 Time-frequency images computed using CWT at 50 Hz.

TFR computed at frequency = 15

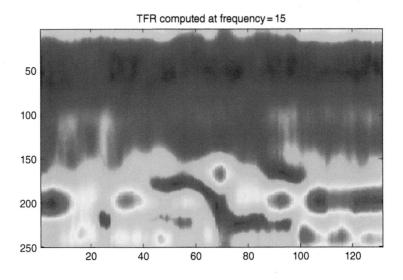

Figure 6.10 Time-frequency images computed using ST at 15 Hz.

the surrounding rocks. It is found that the reservoir resolves better at the frequency range from 20 to 30 Hz. After that, the reservoir frequency response vanishes. The synchron-squeezed transforms did not show interesting results. This might be due to the significant nonstationarity of the seismic traces.

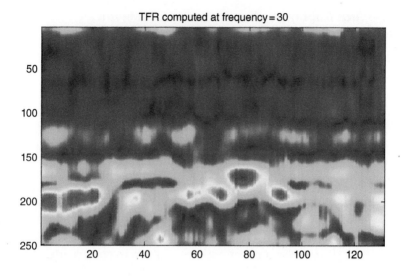

Figure 6.11 Time-frequency images computed using ST at 30 Hz.

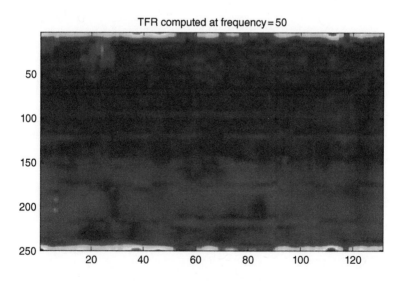

Figure 6.12 Time-frequency images computed using ST at 50 Hz.

6.4. Spectral Decomposition in Modern Reservoir Characterization: A Case Study

Spectral decomposing becomes a routine application in modern seismic interpretation and reservoir characterization. In this section we use OpendTect to carry out time-frequency analysis on challenging seismic data from south Texas. OpendTect is free software intended for seismic interpretation. Full functionality of the program requires acquiring commercial or academic license. However, spectral decomposition tools are provided free of charge. To decompose seismic data, the user can use STFT and CWT; in the CWT three mother wavelets are provided, namely, Gaussian, Mexican hat, and Morlet. The study is performed on a portion of Stratton Field in Kleberg and Nueces Counties of south Texas. The stratigraphic interval we study is the Oligocene Frio Formation, a thick, fluvially deposited sand shale sequence that has been a prolific gas producer in Stratton Field and in several other fields along the FR-4 depositional trend (Figure 6.13).

The seismic interpretation at Stratton field was particularly challenging because most reservoirs, including F39, are very thin beds (less than 5 m), and they are closely stacked, separated only 3 to 5 m vertically in some areas [*Hardage et al.*, 1994]. The dataset is only a 1 × 2 miles subset of a much larger 3-D survey

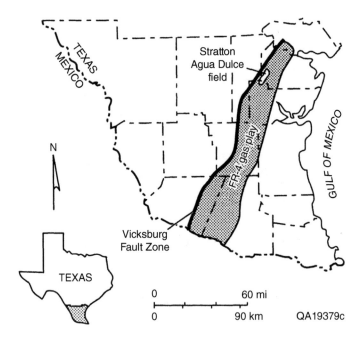

Figure 6.13 Map of the prolific Frio FR-4 gas trend in south Texas showing the location of Stratton Field. (Source: *Hardage et al.*, 1994.)

with a grid of 110 ft × 55 ft (34 m × 17 m) stacking bins, in which a stacking fold of 20 existed over most of the imaging area.

In exploration geophysics, the generally accepted threshold for vertical resolution of a layer is a quarter of the dominant wavelength. A layer is called a thin layer when $1 < \lambda/d \leq 4$, and an ultra-thin layer when $\lambda/d > 4$, where λ is the dominant wavelength within the layer and d is the layer thickness [*Liu and Schmitt*, 2003]. Therefore, calibration with well data and careful seismic to well tie is critical. *Francis* [2005] noted that these gas sand reservoirs are characterized by their high acoustic impedance compared to the surrounding shales, and they typically exhibit a high velocity of 12,000 ft/s. Our objective is to image the F39 reservoir by decomposing seismic broadband into its constituent frequencies and extract information that could not be derived from seismic using other attributes. We first try to address a shallower channel that seismic indicated its expressions. Then we investigate our target reservoir for which seismic failed to detect most of its stratigraphic features.

A careful inspection of seismic data revealed that expressions of a meandering channel were observed at 840 ms (Figure 6.14).

This channel was also documented in *Sinha et al.* [2005]. Spectral decomposition was expected to reveal stratigraphic features of the channel that could not be seen clearly in seismic images. To accomplish this, different frequencies were calculated for a single time slice at this interval. The channel does not resolve at the dominant frequency. At 30 Hz the channel appears clearly, meaning that this frequency is closer to the frequency at which the channel tunes. Similarity, the window-based attribute, was also computed and could successfully define the channel with the same time gate used in short-time windowed Fourier transform (STWFT).

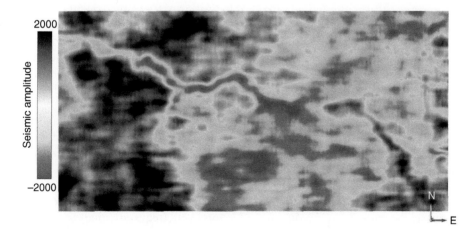

Figure 6.14 Seismic time slice at channel interval shows seismic channel-like expressions.

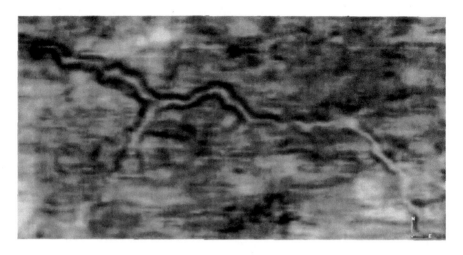

Figure 6.15 Similarity attribute blended with 30 Hz frequency component time slice. Note how the channel is clear.

Then the two attributes were color-blended to improve the channel image. In Figure 6.15, a time slice displays a clear image of the channel.

The main objective to investigate here was the thin reservoir (F39), a gas sand reservoir located at deeper interval (800 ms below the shallow channel). Figure 6.16 shows how the F39 reservoir (in red) is deep (over 6500 ft) and thin.

Seismic data interpretation in *Hardage et al.* [1994] suggested that the reservoir is likely to be subdivided into different compartments. The amplitude trending north-south was interpreted to be clear indication of the compartmentalization. The interpretation proposed by the authors was supported by pressure measurements at the wells W175, W75, and W202 (Figure 6.17). However, seismic images have failed to provide an indication for the two wells W175 and W202. The latter wells are only 200 ft (60 m) apart. Thus, the authors investigated vertical seismic profile data acquired at well W175 instead, to find any information about the reservoir compartmentalization.

Seismic frequency response of subsurface formations is a characteristic that can be attributed to a combination of their rock properties and fluid content. We thus expected that spectrally decomposing the seismic broadband into its individual components might help us reveal some stratigraphic features of the F39 gas sand reservoir that could not be revealed previously.

The same spectral analysis and decomposition workflow as above was executed at the targeted reservoir interval. Dominant frequency analysis demonstrated that dominant frequency of the survey in this zone is around 25 Hz. This decrease in dominant frequency from 55 Hz, in the shallower part, to 25 Hz recorded in the reservoir zone is associated with the attenuation and absorption

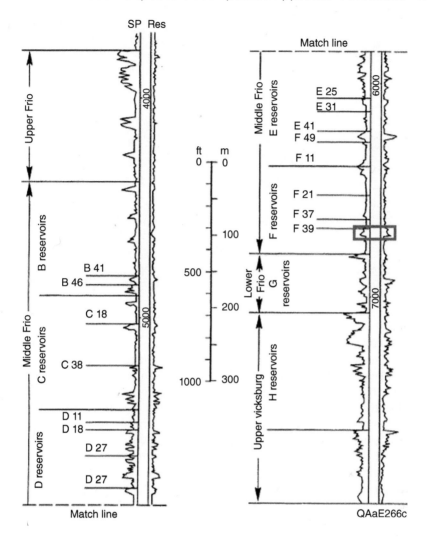

Figure 6.16 Lithological responses obtained from analysis of spontaneous potential and resistivity logs (source: Levey et al., 1994). Note how F39 reservoir is deep (over 6500 ft) and thin (red).

that affect higher frequencies while traveling to deeper formations. Several frequencies were calculated for a time slice passing through the reservoir. Unfortunately, STWFT could not yield significant details about the reservoir.

Seismic signals from this kind of reservoir are composite and result from constructive or destructive interferences of the waves reflected from neighboring reflectors. Consequently, interferences are expected to affect seismic attribute

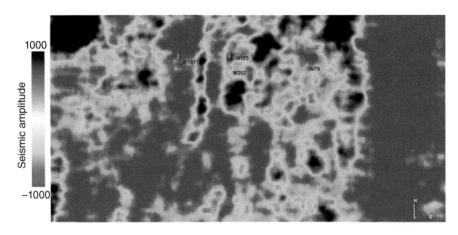

Figure 6.17 Seismic time slice passing through the reservoir shows an amplitude variation between wells. Note that the amplitude change (blue) separating W197 and W75 from other wells was interpreted to be a depositional barrier that resulted in pressure change between wells.

computations, particularly, seismic attributes sensitive to waveform changes and window-based calculation attributes. Spectral decomposition is known to be a good indicator of frequency changes; however, the closely stacked reservoirs' presence makes window-based attribute computation unfavorable. In fact, this would lead to a sum of different formations within the same time window that would result in erroneous interpretations. Thus, continuous wavelet transform-based decomposition was selected to solve the problem. CWT is also sensitive to amplitude changes; it yields spectral image for seismic amplitude at each single time sample. Therefore, errors in imaging and interpreting reservoir frequency behavior would reduce considerably.

We examined the 3-D data using CWT. Interestingly, as Figure 6.18 shows, CWT not only resolved the expected amplitude changes but also illuminated a reservoir meandering channel trending north-south, which was unseen in seismic data. As the 20 Hz slice displays, clear spectrum amplitude changes are observed between wells. These changes are interpreted to be reservoir compartments boundaries. Note that the clear anomalous amplitude that the reservoir compartments and channel show at 20 Hz can be attributed to the result of both thin bed tuning and hydrocarbon charge, which is observed along the channel and around the wells locations as well. The gas charge presence makes the reservoir compartments' reflectivity coefficients larger than those in the adjacent nonhydrocarbon-filled areas. The thin bed tuning effect of those large reflection coefficients preferentially reflects 20 Hz frequency, thus making the gas sand compartments brighter and clearer than at other frequencies.

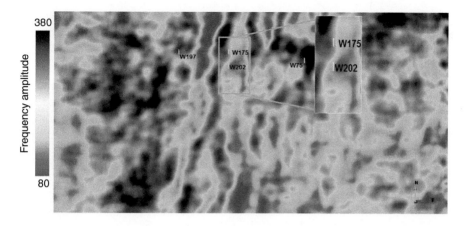

Figure 6.18 20 Hz slice shows clearly compartment boundaries present between wells, causing pressure changes. Note that well W175 is located inside the channel, whereas W202 is found outside the channel.

The frequency component image shows also, unlike seismic, distinct frequency responses at wells W175 and W202. While W175 is penetrating the meandering channel reservoir, W202 is found beyond but near to the channel. This can be due to the fact that the reservoir thickness changes from W202 to W175 and that they penetrate the reservoir at different depth levels, which was confirmed through well information.

6.5. Discussions and Conclusion

It is readily seen that every time-frequency analysis becomes an innovative and promising field in geophysics. This can justify the number of the techniques being developed and adopted to interpret frequency behavior of seismic data. Although these techniques have demonstrated promising results, they do have several drawbacks. The geophysicist must decide which one is most appropriate for the data he or she addresses. In addition, all the methods share the fact that the frequency responses and their interpretations are not unique, so other data (e.g., well data, other seismic attributes) must be incorporated to overcome this problem. They also share the simplicity and easy-to-use and implement on computer programs. As a result, algorithm and source codes become readily available.

It is worth mentioning that most of the techniques are still widely used in today seismic interpretation and reservoir characterization workflows. Successful studies are continuously published and reporting the usefulness of decomposing the seismic signal into its constituent components. Most modern software packages of seismic interpretation have adopted spectral decomposition algorithms in their interpretation tools.

References

Allen, J. B. (1977), Short term spectral analysis, synthetic and modification by discrete Fourier transform, *IEEE Transactions on Acoustic, Speech, Signal Processing*, 25, 235–238.

Auger, F., P. Flandrin, P. Gonçalvès, and O. Lemoine (1996), Time-Frequency Toolbox, CNRS (France) and Rice University (USA).

Chakraborty, A., and D. Okaya (1995), Frequency-time decomposition of seismic data using wavelet-based methods. *Geophysics*, 60, 6, 1906–1916.

Daubechies, I. (1988), Orthonormal bases of compactly supported wavelets. *Communication on Pure and Applied Mathematics*, 41, 7, 909–996.

Daubechies, I., J. Lu, and H. T. Wu (2011), Synchrosqueezed wavelet transforms: An empirical mode decomposition-like tool. *Applied and Computational Harmonic Analysis*, 30, 243–261.

Francis, A. (2005), Limitations of Deterministic and Advantages of Stochastic Seismic Inversion. *CSEG RECORDER*, 2, 5–11.

Goupillaud, P., A. Grossman, and J. Morlet (1984), Cycle-octave and related transforms in seismic signal analysis. *Geoexploration*, 23(1), 85–102.

Han, J., and M. Van der Baan (2013), Empirical mode decomposition for seismic time-frequency analysis. *Geophysics*, 78, 2, O9–O19.

Hardage, B. A., R. A. Levey, V. Pendleton, J. Simmons, and R. Edson (1994), A 3-D seismic case history evaluating fluvially deposited thin-bed reservoirs in a gas-producing property. *Geophysics*, 59, 1650–1665.

Herrera, R. H., J. Han, J., and M. van der Baan (2014), *Geophysics*, 79, V55–V64.

Huang, N., Z. Shen, S. Long, et al. (1998), The empirical mode decomposition method and the Hilbert spectrum for non-stationary time series analysis. *Proc. Roy. Soc. London*, 454A, 903–995.

Levey, R. A., B. A. Hardage, R. Edson, and V. Pendleton (1994), *Fluvial reservoir systems Stratton Field, South Texas*. The University of Texas at Austin, Bureau of Economic Geology, 30 pp.

Liu, J., and K. J. Marfurt (2007), Instantaneous spectral attributes to detect channels. *Geophysics*, 72, P23–P31.

Liu, Y., and S. Fomel (2013), Seismic data analysis using local time-frequency decomposition, *Geophysical Prospecting*, 61, 516–525.

Liu, Y., and D. R. Schmitt (2003), Amplitude and AVO responses of a single thin bed, *Geophysics*, 68 (4), 1161–1168.

Mallat, S. G. (1989), A theory for multiresolution signal decomposition: The wavelet representation. *IEEE Transactions*, 11, 674–693.

Mallat, S. (2009), *A wavelet tour of signal processing*, 3rd ed. Academic Press, Burlington.

Morlet, J., G. Arens, I. Forgeau, and D. Giard (1982), Wave propagation and sampling theory: Part I. *Complex signal and scattering in multilayered media, Geophysics*, 47, 203.

Partyka, G., J. Gridley, J., and J. A. Lopez (1999), Interpretational applications of spectral decomposition in reservoir characterization. *The Leading Edge*, 18, 353–360.

Sinha, S., P. Routh, P. Anno, and J. Castagna (2005), Spectral decomposition of seismic data with continuous-wavelet transform. *Geophysics*, 70(6), P19–P25.

Stockwell, R. G., L. Mansinha, and R. P. Lowe (1996), Localization of the complex spectrum: The S transform. *IEEE Transactions on Signal Processing*, 44, 998–1001.

Thakur, G., E. Brevdo, N. S. Fuckar, and H.-T. Wu (2013), The synchrosqueezing algorithm for time-varying spectral analysis: Robustness properties and new paleoclimate applications. *Signal Processing*, 93, 5, 1079–1094.

Torres, M., M. Colominas, G. Schlotthauer, and P. Flandrin (2011), A complete ensemble empirical mode decomposition with adaptive noise. *IEEE International Conference on Acoustics, Speech and Signal Processing*, 4144–4147.

Wang, Y. H. (2007), Seismic time-frequency spectral decomposition by matching pursuit. *Geophysics*, 72(1), 13–21.

Wigner, W. (1932), On the quantum correction for thermodynamic equilibrium. *Physical Review*, 40, 749–759.

Wu, Z., and N. E. Huang (2009), Ensemble empirical mode decomposition: A noise assisted data analysis method. *Advances in Adaptive Data Analysis*, 1, 1–41.

7. ELECTROMAGNETIC GEOPHYSICAL RESEARCH WITH CONTROLLED SOURCE

Olga Hachay[1] and Oleg Khachay[2]

Abstract

New ideas of the 20th and 21st centuries are considered. According to these ideas, new theoretical and methodical approaches are examined for detailed investigations of the structure and state of the geological medium and its behavior as a dynamic system in reaction to external man-made influences. To solve this problem, it is necessary to use geophysical methods that have sufficient resolution and that are built on more complicated models than layered or layered-block models. One of these methods is the electromagnetic induction frequency-geometrical method with controlled sources. Here we consider new approaches using this method for monitoring rock shock media by means of natural experiments and interpretation of the practical results. This method can be used on oil production in mines, where the same events of nonstability can occur.

7.1. Introduction

This overview is based on the key ideas of geophysics of the 21st century, taking the point of view of geologist and academician *A. N. Dmitrievskiy* [2009]:

The geophysics of the 20th century is the understanding of features such as: geophysical fields which are indicators of the processes that occur in the lithosphere; geophysical parameters which are remotely measured or are closely correlated with the material-structural characteristics of the geological medium (on the micro- and macroscales); analysis of space-time and the energetic distribution of the geophysical field can give information about the space-time distribution of the properties of the geological medium; registration and analysis of the geophysical field through monitoring can give information about geodynamical processes in the lithosphere and in deeper layers.

[1]*Institute of Geophysics Ural Branch of Russian Academy of Sciences (UB RAS), Ekaterinburg, Russia*
[2]*Ural Federal University, Ekaterinburg, Russia*

Oil and Gas Exploration: Methods and Application, Monograph Number 72,
First Edition. Edited by Said Gaci and Olga Hachay.

The practical problems of geophysics in the 20th century were a powerful incentive for the evolution of theoretical and experimental physics of thin-layered, porous, and crack media. As a result, new classes of mathematical models of fluid-saturated heterogeneous media were developed, and different physical and physical-chemical effects that occur on the "solid skeleton-liquid" boundaries were revealed. Geophysics for the first time raised a question about the possibility of constructing physical-geological and mathematical models of geological objects and processes.

The applied geophysics of the 20th century realized the possibility of researching the same geological objects on the microlevel (nuclear geophysics), mesolevel (electrical, thermal, magnetic and acoustical fields), and macrolevel (field of elastic waves and low-frequency electromagnetic fields).

Geophysics allowed us to answer the principal questions for geologists: what is the depth and geometry (sometimes the morphology) of the researched geological object? What is the material-structural content of the geological object? Where and how are the subvertical and subhorizontal heterogeneities and primarily the zones of macrocracks located? What are the filtration conditions for the fluid-porous aggregate? What are the thermal and dynamic conditions of the object sought?

7.2. Theory and Theoretical Methods of Solving Complicated Problems

"Taken separately, each geophysical method is linked with some uncertainty of the geological interpretation of the results of observations. This uncertainty is decreased and the probability of correct interpretation increased with the use of certain methods. Therefore, there naturally arises an aspiration to use integrated geophysical methods" [*Bulashevich*, 1950]. What does it mean to speak of the "geological interpretation" of the results of geophysical observations [*Hachay*, 2002]? It means that, besides the spatial distribution of the physical properties of the geological medium, which is defined in the frame of the influence on the medium of one field or another, or registering one or another emission of the medium, it is important to define the modern material content and the state of the geological medium [*Bulashevich and Ermakov*, 1972; *Bulashevich et al.*, 1980]. This is due to the difficulty of interpreting a complex constructed geological model and researching the mechanisms that lead to its nonstationary reconstruction [*Hachay*, 2000]. Thus, it is in contrast to the problems, which we solved earlier [*Hachay*, 2002], linked with mapping the 3-D layered-block medium. Integration of the interpretation model was achieved by overlapping separate models that we constructed based on the independent interpretation of different types of geophysical field data: gravitational, magnetic, seismic, electromagnetic, and thermal. Here we must take into account the thermodynamic relationship of the material parameters of the medium. So the interpretation process is no longer an independent solution of inverse problems for each field separately. The operator of the inverse problem in the case of complex interpretation consists of a system of equations for fields that are linked with thermodynamic material relations. As examples, we can consider the algorithms defining the mechanisms of convection of the Earth's mantle, and the interpretation of the anomalies of the natural field, linked with electro-kinetic events [Hachay and Khachay, 1993; Goldin, 2002].

The theory of interpretation of geophysical fields is closely linked with the solution of incorrect problems and with regularization theory, which was developed by A. N. Tikhonov, V. K. Ivanov, and V. V. Lavrentiev and the scientific schools they created. As a result of its application, significant results were achieved, enabling the development of local, regional, and global distributions of the physical properties of the Earth, Moon, and other planets, on which geophysical research has been provided in the frame of partly homogeneous models and of linear approximation of the distribution of fields.

The subsequent evolution of geophysical methods for the research of more complicated media and events necessarily leads to more complicated problems of mathematical geophysics, for which not only inverse but also direct problems become substantially incorrect [Mazurov, 1990; Samarsky, 1989]. In connection with this, on one hand it is necessary to develop a new conception of regularization for the simultaneous solution of a system of different types of operator equations. On the other hand, another approach is possible: creating equal types of interpretation algorithms on the basis of the solution of equal types of operator equations, which differ in their core functions.

Using the second approach and the ideas of A.V. Tsirulskiy about the two-stage interpretation of potential fields, we developed a single conception of a three-stage interpretation of electromagnetic and seismic fields in a dynamic approach [Hachay, 1996]. In that conception, for regularization, we used the idea of filtering the input data in the area of a 1-D operator of inverse problem solution definition and the idea of approximating anomaly fields by the class of fields with given properties in advance [Hachay, 1999]. We developed algorithms that realize all stages of the interpretation, which allow us to investigate the application of that approach for analysis of practical data [Hachay and Novgorodova, 1997; Hachay, Bodin et al., 1998]. For simultaneous use of the data of different fields, we developed the main principles of constructing area systems of observation, which allow us on one hand to obtain agreement with the normal field of the data base, and on the other hand, to realize a 3-D interpretation in the frame of unified algorithms [Hachay and Hinkina, 1997]. Examples of their practical use testify that the suggested integration is highly informative [Hachay, 1980, 1991; Hachay, Novgorodova et al., 1999a; Hachay, Druzhinin, et al., 1999].

Let us consider the components of that conception:

1. In the first stage, electromagnetic and seismic parameters of the imbedded elastic and geoelectrical heterogeneities of the horizontal layered-block medium are defined.

2. In the second stage, we realize the fitting of the anomaly field by a field of singular sources immersed into the medium (with physical parameters defined in the first stages), which are equivalent by the field to local geoelectrical and elastic heterogeneities. Thus, we can define the geometric model of apparent local heterogeneities or group heterogeneities and their respective locations within the layered-block imbedded medium.

3. In the third stage, we define the surfaces of the heterogeneities being sought in relation to the values of the physical parameters of anomalous forming objects.

7.3. Practical Realization of the New Conception

For the practical realization of these conceptions, we developed the following theoretical and practical results [*Hachay and Novgorodova*, 1997; *Hachay, Novgorodova, et al.*, 1998]:

1. A single system of observation was tested and developed for alternating electromagnetic and seismic fields with the use of a locally controlled source of excitation. The choice of the source of excitation is thus defined: (a) single geometry of the normal field, (b) by the absence of one or more components in the measuring field for the case of the quasi-layered medium. These properties in the case of the electromagnetic field have as a source a vertical magnetic dipole, and in the case of the seismic field, a vertically acting force. The existence of a local source of excitation allows us, from a given set of observations, to regularly realize an overlapping by different angles of observation (heterogeneity of the source of excitation). As the input data base for interpretation, we use three components of the magnetic field and three components of the field of elastic displacement as functions of space coordinates and time or frequency. In the case of surface observations, the data are fixed on the surface for a set of distances between the source and the receiver as a function of time or frequency. Analysis of the solution of direct problems for the seismic and electromagnetic cases shows [*Hachay and Hinkina*, 1997] that the single calculation approach can be realized for both the field data in the case of preliminary processing of electromagnetic data on the real axes, and seismic data on the image axes of the complex frequency plane. Therefore, the whole further process of interpretation needs to be organized on that plane and not to transit to the time area.

2. For the realization of the second stage, that is, to achieve the analysis of the anomaly field, modules of horizontal components of seismic and magnetic field as functions of space coordinates and a real or imaginary parameter of frequency were introduced. For the cylindrical coordinate system, the ratio of the φ component to the ρ component is used, which has a sense of parameters of seismic and geoelectrical heterogeneity and quantitatively characterizes

the degree of medium deviation from the horizontal-layered model. The fitting of these parameters is achieved by construction of an approximation [*Hachay and Novgorodova*, 1997; *Hachay, Bodin, et al.* 1998], which is done on the basis of an explicit expression of fields of singular sources as follows: for the seismic field, it is a set of point force sources that act on a section of finite length in an arbitrary direction; for the electromagnetic field, we have a set of current lines of finite length. Analysis of the direct problem for the seismic case with the influence of the source located in the horizontal plane allows us to conclude that to achieve morphology unity for the fitted anomalous fields it is preferable for the electromagnetic field to use the system of singular sources as closed-current contour-horizontal magnetic dipoles. In spite of that, each of these systems is complete and can be used for fitting anomalous electromagnetic fields [*Hachay and Hinkina*, 1997], through integrated and coupled interpretation of the seismic and electromagnetic fields. In addition to changing the type of construction, it is necessary to include for the electromagnetic case an additional parameter, which is defined as the ratio of electrical horizontal components ρ and φ, whose space distribution is closely linked with the distribution of the parameter of seismic heterogeneity in its shear part. Therefore, for the electromagnetic case it is necessary to measure additional horizontal electric components or corresponding derivatives of the magnetic field.

We have developed algorithms that allow us to research the possibility of using this approach for the interpretation of practical data [*Hachay*, 1980, 1991]. For the simultaneous use of different field data, we developed new main principles for constructing area systems of observations; these allow us on one hand to obtain a match with the normal field data base, and on the other hand, to realize a 3-D interpretation in the frame of uniform algorithms [*Hachay and Novgorodova*, 1997]. The examples of practical use testify to the highly informative nature of the suggested integration [*Hachay, Novgorodova et al.*, 1999a; *Hachay, Druzhinin et al.*, 1999]. At the Institute of Geophysics Ural Branch of Russian Academy of Sciences (UB RAS), a device was developed by Professor Chelovechkov that was used for near-surface induction electromagnetic research and also a seismic device called "Sinus," developed by Professor Senin. In the frame of the project, which was approved by the Ministry of Natural Resources together with the Ural Committee of Natural Resources, the abovementioned method together with the method of processing and interpretation has been effectively used to provide prospecting works for updating of the structure of the platinum placer on the flank of the Lobva deposit. The forecast geophysical sections and models were confirmed by borehole data and overburden operations.

The "planshet" electromagnetic method was used for a solution of the problem of tracing serpentinites in the Ural region (grant RFBR No. 99-05 64586). The integrated seismic and electromagnetic method, adapted for research on the structure and state of the rock massif, is used in an underground variant as a volume system of observation, realized for the first time in mine conditions (grant RFBR No. 99-05-64371).

The global and west Siberian experience of oil and gas reservoir engineering quite definitely testifies to the significant localization of the oil and gas zones in the structure of the deposits. It is evident that without highly detailed and solid geological and geophysical information about the spatial location of the axis parts of sub-vertical destruction zones (anomalous concentration of crack zones in the Earth's crust on different age levels) on the outworking deposits, we cannot look forward to a highly economical realization of oil mining [Hachay, Bodin, et al., 2001].

The developed method allows adjustments to be made to the solution of highly complicated geological problems. Its use allows us to perform the system of observation in such a way as to both achieve a flexible reconstructive observation detail and organize the input database to be much closer to the domain of defining the inverse problem operator in the class of layered-block models with hierarchic inclusions. With the use of regularization methods, this allows us to obtain solutions in the form of equivalent models, which are near to reality. The concrete possibilities of that realization are defined by the technical side: the capacity of the excitation source and the sensitivity of the receiving system. In the case considered, this method helped to research the changed forms of the kimberlitic bodies that occur during the evolution of diamond zones. The method was verified by geological surveys in different geological conditions [Hachay and Novgorodova, 1999; Hachay, Druzhinin, et al., 1999; Hachay, Novgorodova, et al., 2000, 2003a; Hachay, 1997; Hachay, Bodin, et al., 1997; Khachay, 1998; Hachay, Khachay, et al., 1998]. Realization of the method for lower excited and measured frequencies of alternating magnetic field by fixed detailing can make it possible to map the structure of the oil gas deposit and to zoom the seismic information in order to assign industrial boreholes, along with a sharp reduction in the volumes of additional preliminary drilling. The developed planshet induction method using a controlled source of excitation has been included in the complex of electromagnetic methods used in Egypt to solve different problems of near-surface geophysics and archeology [Atya, Khachay, et al., 2010a and b].

7.4. Problem of Monitoring Active Zones of Geological Medium Using Controlled Sources

One of the fundamental problems of mining, which is traditionally related to the problems of geomechanics, is creating a theory and methods for research of the structure and state of the rock massif to allow forecasting and prevention of catastrophic events by working out the deposit. The basic idea of developing an integrated geophysical and geomechanical approach in the Ural was inspired by Professor N. P. Vloch [1994, 1997].

Estimation of the massif state is a more complicated problem than mapping its structure. The state is a sum of the stress and phase state, which can lead to either elastic or plastic massif deformations. An interrelation exists between these two types of state, so in research on the massif we must consider them together, because for each of them, a sharp change of the corresponding state can lead to the loss of massif stability and rock burst. So, the stress state depends on the initial state of the stress field, which is defined by the geological structure and its physical and mechanical properties: fracturing and porosity. The phase state is defined by the mobility of microstructure elements of the rock massif and its many phases: that is, by the influence not only of the skeleton, but also of the liquid and gaseous phase, which saturates the cracks and pores of the massif. Its changes are defined by man-made influences on the massif, such as man-made karsts, and by the pumping of mechanical energy during the mass explosions that are needed for outworking.

The phenomenon of the nonstationary rock massif is a known factor today [*Adushkin and Tsvetkov, 1997; Shkuratnik and Lavrov, 1997*]. It may be barely noticeable as increasing rock cracks and may also be registered by dynamic events with initially small energy and then with ever higher energy [*Catalogue of Rock Bursts, 1986*]. The latter relate to catastrophic events, which are initiated by interior and man-made external causes. To forecast these events by working out the specific ore deposit, it is important to have data about the degree of the rock massif's relaxation under a controlled mechanical influence. Thus, the problem is related to the following:
• mapping and identification zones: inclusions in the layered-block isotropic medium by physical properties of the massif and tracing their migration under the influence of changing man-made stresses
• estimation of the massif state from the type of inclusions: contact on the borders with rocks of different material content, or a crack medium with a different degree of water saturation and its fixed changing in time.

All the listed features make the problem of creating an integrated approach to researching the rock massif state based on reciprocally refined geophysical and geomechanical methods ever more pressing.

Using the results of theoretical analysis of the possibility of using geophysical fields of different natures and the practical use of new methods of rock massif research, we formulated the main principles of the currently realized system for monitoring the rock massif's stress state. In this way, the rock massif is presented as a set of local inclusions with a hierarchic structure, which are embedded in the layered-block isotropic medium. Monitoring is conducted for research of the nonstationary appearance of these local zones, linked with the changing of the massif state.

The developed monitoring system is organized as three related cycles: [3]-[2]-[1], where the first cycle (inner) is constructed on the basis of the iteration principle and consists of the following procedures:
a.   Revealing local heterogeneous zones by means of their physical properties; conductivity is provided by using the volume method of induction electromagnetic research, developed in the Institute of Geophysics UB RAS.

As a result, we create a geometrical model of the structure of the researched rock massif area [*Hachay, Novgorodova, et al.*, 1998];

b.   Providing mathematical modeling of the seismic field in the frame of the given geometry of the medium with different elastic characteristics [*Hachay*, 2003; *Vloch, Lipin, et al.*, 1994];

c.   Constructing an integrated electromagnetic and seismic model of the rock massif's local zones;

d.   Calculating the tensor components of secondary stresses that are created by heterogeneous zones by electric and elastic properties;

e.   Constructing a combined model using geomechanical and geophysical data of the stress state in the area of the local heterogeneous zone;

f.   Making a decision to return to point a, or to shift to the second cycle, which is defined by the criterion of fulfilling point e.

In the second cycle, we separate and classify the results of the monitoring, fitting the physical and mechanical properties of the searched local rock zone with the approaches used [*Vloch*, 1995], and forming a time series of their changes under the influence of man-made and natural factors.

In the third (outer) cycle, we define the quantitative interrelation between the physical and mechanical parameters, which characterize the secondary stress state of each local rock massif zone in dependence on the geological, man-made, and seismogenic factors.

This system was tested on a set of massifs, which differ in their material content, geological classification (sedimentary, volcanogenic-sedimentary, and volcanogenic), and degree of burst danger. Specifically, it was used on the following mines: "Magnezite", SUBR, Tashtagol, Uselgi, Mirniy, and Berezniki. Each of these allows us to reveal the structural peculiarities and massif behavior, and to define and classify the factors that influence the massif's stability [*Hachay, Bodin, et al.*, 1999; *Hachay, Novgorodova, et al.*, 1999b; *Hachay and Novgorodova*, 2000; *Hachay and Khachay*, 2001; *Hachay, Bodin, et al.*, 2001; *Hachay, Vloch, et al.*, 2001].

7.5. Research on the Effect of Self-Organization of Rock Burst Mines in Massifs, Using Active Electromagnetic Monitoring

At the present time, there is sufficient accumulated information to testify to the two most significant peculiarities of the modern evolution of the geological medium: mechanical shifting of the earth's material occurs on arbitrary space and time scales; and the material available for research forms a hierarchical block structure, which is a result of destruction or disintegration processes [*Goldin*, 2002; *Hachay and Khachay*, 2003; *Hachay*, 2004a].

In the understanding of the formation and evolution of the structural levels of deformation in solid bodies, a significant role is played by the theoretical and experimental results obtained from specimens [*Panin, Lichachev, et al.*, 1985]. With these results, a new approach was developed that uses the definitions of dissipative structures in nonequilibrium systems [*Nikolis and Progogine*, 1979], for which there are self-organization processes on each hierarchic level. As shown in the paper by *Nikolis* [1989], self-organization occurs by the presence of the hierarchic structure. This approach can be used for research of such natural and man-made systems in rock massifs that are in the process of being worked out. To describe these, the model of an open dynamic system can be used [*Nikolis and Progogine*, 1990]. Analysis of the self-organization display can provide information on the system stability and contribute to the development of massif state stability criteria as a whole relative to dynamic events of the given energetic class. This overlaps with the statement made in *Panin et al.* [1985] about the hypothesis of medium-scale division.

In the papers [*Hachay*, 2003, 2004b] using active 3-D electromagnetic induction space-time monitoring [*Hachay*, 2006], we were able to determine that the structure of rock massifs with different material content can be described by the model of a hierarchic discrete medium. In the frame of specific modification of the method, we could trace three hierarchic levels. The disintegration zones [*Hachay*, 2007] in the near-hole space are located nonsymmetrically in its bottom and roof, which may be evidence of the nonequilibrium of the system. These zones are discretely located, that is, there are intervals where they are completely absent in the vicinity of the hole. The maximum changes in the massif, which are under man-made influences, simply appear by the changing in time of the morphology of the spatial location (Figure 7.1 a–d). Now we will analyze the changes in parameters.

$$Spint\left(N, T\right) = \sum_{i=1}^{k_N} \tilde{M}_0^i$$

where N is the number of the interval, by which the underground near-hole area is divided: $N = 1$ (0–1 meters), $N = 2$ (1–2 m), $N = 3$ (2–3 m), $N = 4$ (3–4 m), $N = 5$ (4–5 m), $N = 6$ (5–6 m), $N = 7$ (6–7 m), $N = 8$ (7–8 m), $N = 9$ (8–12 m), $N = 10$ (12–17 m); T is cycles of observations during each year: $T = 1$ (2000), $T = 2$ (2001), $T = 3$ (2002), $T = 4$ (2003); k_N is the quantity of revealed heterogeneities in the interval N along the whole length of the hole during one year for each cycle of observation of the massif, which belong to different groups of stability. First, we will analyze the stable massif of ort 4, horizon -210 (Figure 7.2 a–d).

In Figure 7.2 a and b we can see that in 2000, the maximum value of the parameter *SP int* is related to the first interval from 0 to 1 m, while in 2001, the maximum of the parameter *SP int* shifted to the 4th interval. In 2002, the maximum value of the parameter *SP int* shifted deeper into the massif. In 2003, the maximum value of the parameter *SP int* became less and was practically equal for all intervals.

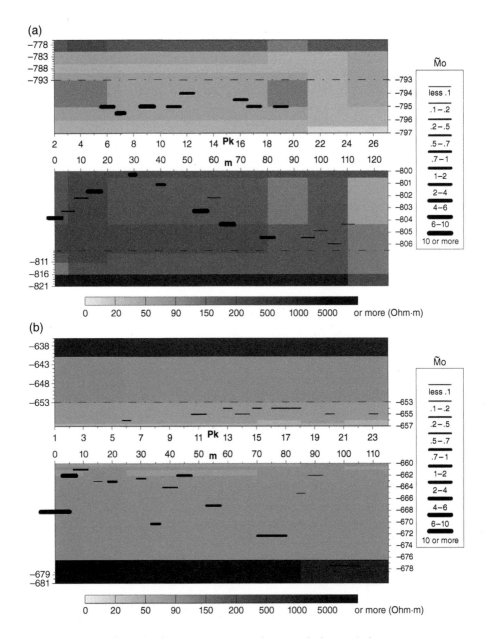

Figure 7.1 Display of self-organization in the morphology of disintegration zones, revealed by the data from active electromagnetic induction monitoring. (a) Geoelectrical section for the roof (top) and the ground (bottom) of the massif ort 19, horizon –350, frequency 20 kHz, observations in 2003. (b) Geoelectrical section for the roof (top) and the ground (bottom) of the massif ort 8, horizon –210, frequency 10 kHz, observations in 2002. (c) Geoelectrical section for the roof (top) and the ground (bottom) of the massif ort 19, horizon –350, frequency 20 kHz, observations in 2002. (d) Geoelectrical section for the roof (top) and the ground (bottom) of the massif ort 8, horizon –210, frequency 10 kHz, observations in 2002.

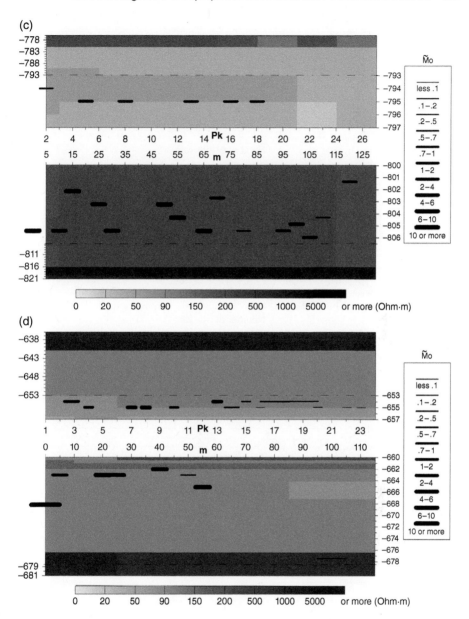

Figure 7.1 (Continued)

In 2004 (Figure 7.2e), the maximum value of the parameter *SP int* shifted again to the contour of the bottom hole. Thus, the results obtained allow us to conclude that there is a cycled shifting of disintegration zones, which we can observe from the location of the maximum value of the parameter *SP int*. For the

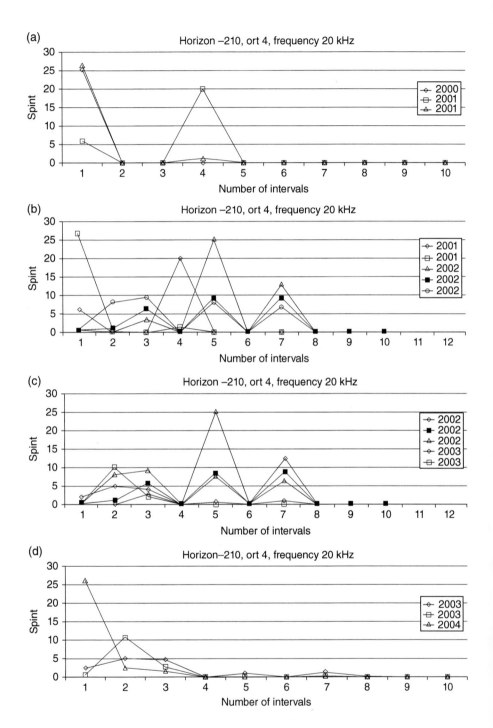

Figure 7.2 Distribution of interval intensity for massifs with different degrees of stability for different orts in the Tashtagol mine.

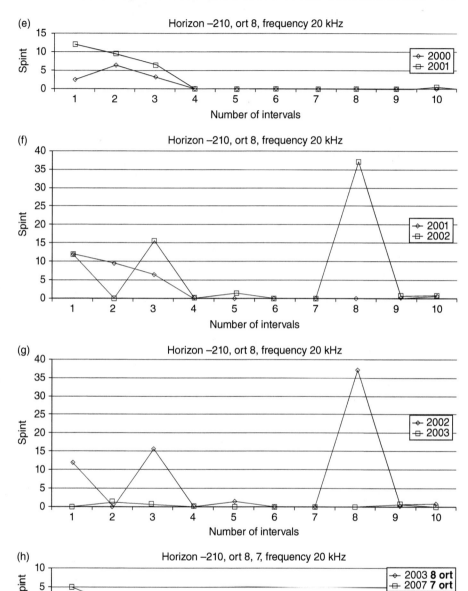

Figure 7.2 (Continued)

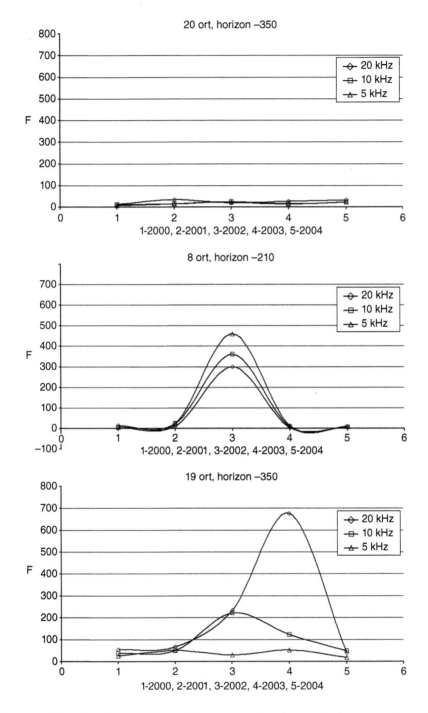

Figure 7.3 Changing the parameter *F*. This characterizes changes in the structural massif, defined from electromagnetic monitoring data. Axis X: Numbers of observation cycles: 1 = August 2000, 2 = August 2001, 3 = August 2002, 4 = August 2003, 5 = July 2004.

frequencies 5 and 1 kHz, an analogous picture exists, but the cycled shifting is removed in one year to the side of lagging. It is of great interest to analyze the dynamics of the parameter $SP\ int(N, T)$ during each year for all five cycles of observation of the massif, which belongs to the nonstable massif group. This is the massif of ort 8, horizon -210 (Figure 7.2 e–h).

Sharp changes in the maximum values of the parameter $SP\ int$ appear in 2002, when it is located very deep, from 8 to 9 m from the hole contour inside the bottom. In 2003, the parameter $SP\ int$ became less sharp and the same value persisted in 2004. The massif was in a stage of hard disintegration after a rock shock.

The dynamics of the distribution of the parameter $SP\ int$ is identical for all frequencies. The registered display of morphology cycling of the distribution of the parameter $SP\ int$ as a function of time characterizes the massif's stability relative to heavy energetic dynamic events, while the reconstructed cycling of the massif structure can occur together with weak dynamic events. Simultaneous analysis of the maximum location (N-number of intervals) of the distribution of parameter $SP\ int$ inside the massif (in the massif bottom) and its value as a relation of parameter $F(T) = SP\ int^{max}(T, N_0) \times N_0(T)$ as a function of the cycle of observation T allows us to establish precise classification boundaries between the massif in a stable state (Figure 7.3a) and the disintegrated massif (Figure 7.3b). So for a stable massif, the value of parameter $F(T)$ is bounded by 100, while for the disintegrated massif the behavior of parameter $F(T)$ is the same for all frequencies and that parameter has a maximum that is three times and more greater than the value 100. In the massif, where structural reconstructions occur near the holes contour and do not involve the deeper parts, the distribution from T of the parameter $F(T)$ differs for different frequencies (Figure 7.3c). As a result, we can determine that the massif is in a state of nonstable evolution.

7.6. Conclusions

The key ideas of 21st-century geophysics from the point of view of geologist academician *A. N. Dmitrievskiy* [2009] are as follows:

The geophysics of the twenty-first century states that: the Earth is a self-developing, self-supporting geo-cybernetic system, in which the role of the driving mechanism is played by the field gradients; the evolution of geological processes is a continuous chain of transformations and the interaction of geophysical fields in the litho-, hydro-, and atmosphere.

The use in geophysical principles of a hierarchic quantum of geophysical space, nonlinear effects, and the effects of reradiating geophysical fields will allow the creation of a new geophysics.

The transfer from research on pure geophysical processes and technologies to geophysical-chemical processes and technologies will allow us to solve the problems of forecasting geo-objects and geo-processes in previously unavailable geological-technological conditions.

The results obtained allow us to make the following conclusions. According to the key ideas of academician Dmitrievskiy, the rock massif is a many-ranked hierarchical structure. Research on the state dynamics, its structure, and the effects of self-organization can be provided by geophysical methods, which are built upon the model of that medium. The use of the planshet multilevel induction electromagnetic method with a controlled source of excitation and a corresponding method of processing and interpretation has allowed us to reveal the disintegration zones, which are indicators of massif stability, and understand the low productivity of oil recovery from boreholes.

Acknowledgments

The work was fulfilled with the support of grants RFBR 10-05-00013-a, 07-05-00149-a, 02-05-64229, and 02-05-96433 projects together with Institute of Mining Siberian Branch of RAS and programs of RAS.

References

Adushkin, A., and V. Tsvetkov (1997), Influence of structure and geodynamics on the stress state of Earth's crust. *Proc. Conf. on Problems of Rock Mechanics*, St Petersburg, pp. 7–12 (in Russian).

Atya, M., O. Khachay, et al. (2010a), Geophysical contribution to evaluate the hydrothermal potentiality in Egypt: Case study: HammamFaraun and Abu Swiera, Sinai, Egypt. *Earth Sciences Research Journal*, 14(1), 44–62.

Atya, M., O. Khachay, et al. (2010b), CSEM imaging of the near surface dynamics and its impact for foundation stability at quarter 27, May 15, Helwan, Egypt. *Earth Sciences Research Journal*, 14(1), 76–87.

Bulashevich, Y. (1950), The relations between electrical and gravitational anomalies. *Proc. Mining-Geological Institute*, issue 19, *UFAN Geophysical Book*, 1, 3–13 (in Russian).

Bulashevich, Y., B. Djakonov, et al. (1980), Methods and results for a joined model of the Ural crust and upper mantle construction. In *Kiev, Tektonosphere of Ukraine and other regions of USSR*, pp. 181–195 (in Russian).

Bulashevich, Y., and N. Ermakov (1972), *Perspectives of use of geophysical methods of regional research of the Ural and searching for minerals, ore base of the Ural*. Moscow, Nauka, pp. 92–109 (in Russian).

Catalogue of rock bursts on ore and non-ore deposits (1986), Leningrad, VNIMI (in Russian).

Dmitrievskiy, A. (2009), *Selected works*. Institute of Oil and Gas Problems of RAS, Moscow, Nauka, vol. 2 (in Russian).

Goldin, S. (2002), Destruction of lithosphere and physical mesomechanics. *Physical Mesomechanics*, 5(5), 5–22 (in Russian).

Hachay, O. (1980), Uniform method of inverse problem solution of electromagnetic soundings into 1-D medium. *Izv. AN USSR, PhysikaZemli*, 5, 51–60 (in Russian).

Hachay, O. (1991), About the inverse problem solution for 3-D alternating electromagnetic fields. *Izv. AN USSR, FizikaZemli*, 6, 50–57 (in Russian).

Hachay, O. (1996), Three-staged method of interpretation of alternating electromagnetic fields and its practical realization. In *Electromagnetic Research with Controlled Sources*. St. Petersburg, RAS, pp. 30–31 (in Russian).

Hachay, O. (1997), Three-staged concept of common interpretation for 3-D electromagnetic and seismic fields and some results of its practical realization. In *Engineering and Environmental Geophysics for the 21st Century*. China, Chengdu, pp. 286–292.

Hachay, O. (1999), Theoretical principles of integrating interpretation of geophysical fields on the base of the method of inverse problem solution with use of regularization methods. *Proc. 1st Russian Conference Geophysics and Mathematics*, Moscow, pp. 229–231 (in Russian).

Hachay, O. (2000), Integrated geophysical research (theory and practical results). *Ural Geophysical Bulletin, IGF UB RAS*, 1, 107–110 (in Russian).

Hachay, O. (2002), Ideas of Y. P. Bulashevich for modern understanding of the problem of integrating geophysical methods. *Ural Geophysical Bulletin, IGF UB RAS*, 3, 42–46 (in Russian).

Hachay, O. (2003), About the question of research on the structure and state of geological heterogenic medium and their dynamics in the frame of discrete and hierarchic model. *Geomechanics in Mining*. Ekaterinburg, IM UB RAS, pp. 30–38 (in Russian).

Hachay, O. (2004a), On the question of the structure and state of geological heterogeneous non-stationary medium in a frame of discrete hierarchic model. *Russian Geophysical Journal*, 33-34, 32–37 (in Russian).

Hachay, O. (2004b), Phenomenon of self-organization in rock massif by man-made influence. *Physical Mesomechanics*, 7(Special issue), 292–295 (in Russian).

Hachay, O. (2006), The problem of the transition process research of stress and phase state between heavy man-made explosions. *Mining Information Analytical Bulletin, Moscow, MSSU*, 5, 109–115 (in Russian).

Hachay, O. (2007), Geophysical monitoring of the massif state with use of the paradigm of physical mesomechanics. *FizikaZemli*, 4, 58–64 (in Russian).

Hachay, O., V. Bodin, et al. (1997), A new complex near-surface electromagnetic and seismic technique for 3-D research of heterogeneous and non-stationary medium. *Engineering and Environmental Geophysics for the 21st century*. China, Chengdu, pp. 181–189.

Hachay, O., V. Bodin, et al. (1998), Theoretical principles of frequency-geometric 3-D research of mining-geological medium. *Proc. Int. Conf. Rock Geophysics*, St. Petersburg, pp. 583–590 (in Russian).

Hachay, O., V. Bodin, et al. (1999), Combined approach to interpretation data of 3-D seismic and electromagnetic fields in frequency-geometric variant by local excitation. *Questions of Theory and Practice of Interpretation Gravi-magnetic and Electrical Fields*. Ekaterinburg, pp. 68–69 (in Russian).

Hachay, O., V. Bodin, et al. (2001), Method of mapping zones of potential non-stability of rock massif of different material content with use of data of dynamic seismic and electromagnetic induction research. *Mining Information Analytical Bulletin, MSSU, Moscow*, 2, 10–16 (in Russian).

Hachay, O., V. Druzhinin, et al. (1999), Evolution of Earth's crust 3-D heterogeneities mapping methods on the base of seismic frequency-geometric research. *Proc. Int. Conf. 50 years of DSS*, Moscow, p. 124 (in Russian).

Hachay, O., and T. Hinkina (1997), About an algorithm of inverse problem solution for the seismic problem of elastic 1-D medium. In *Astronomic–Geodetic Research*, UrGU, Ekaterinburg, pp. 174–178 (in Russian).

Hachay, O., A. Khachay, et al. (1998), New geophysical technique for permafrost research in the Polar Ural. *Proc. 7th Int. Conf. Permafrost,* Yellowknife, N.W.T., Canada, 23–27 June 1998, Abstract, 118.

Hachay, O., and M. Khachay (2003), New approaches to analysis of highly complicated non-stationary media. *Ural Geophysical Bulletin.* IGF UB RAS, Ekaterinburg, 1, 24–28 (in Russian).

Hachay, O., and O. Khachay (2001), Method of defining an equation of factor interrelation, which describes rock shocks (using data from SUBR). *Mining Information Analytical Bulletin*, MSSU, Moscow, 2, 46–49 (in Russian).

Hachay, O., and Y. Khachay (1993), About the identification of physical mechanisms of mantle convection on the base of the method of inverse problem. *Geology and Geophysics*, 6, 15–21 (in Russian).

Hachay, O., and E. Novgorodova (1997), Experience of area induction research of sharp heterogeneous geoelectrical media. *FizikaZemli*, 5, 60–64 (in Russian).

Hachay, O., and E. Novgorodova (1999), Use of the new 3-D method of electromagnetic research of the structure of rock massive. *FizikaZemli*, 5, 7–12 (in Russian).

Hachay, O., and E. Novgorodova (2000), Mapping and identification of disintegration zones of rock massif of different material content by electromagnetic method. *Institute of Geophysics UB RAS*, Ekaterinburg, pp. 114–123 (in Russian).

Hachay, O., E. Novgorodova, et al. (1998), 3-D electromagnetic research of structure and state of rock massif. *Proc. Int. Conf. Rock Geophysics,* St. Petersburg, pp. 591–598 (in Russian).

Hachay, O., E. Novgorodova, et al. (1999a), About problems of near surface geoelectrics and some results of their solution. *FizikaZemli*, 5, 47–53 (in Russian).

Hachay, O., E. Novgorodova, et al. (1999b), Electromagnetic monitoring of zones of high cracks migration of rock massif by man-made influence. *Geodynamics and Stress State of the Inner Structure of the Earth.* Novosibirsk, SB RAS, pp. 363–367 (in Russian).

Hachay, O., E. Novgorodova, et al. (2000), Mapping of 3-D conductive zones using area systems of observation in the frame of 3-D frequency-geometrical method. *Geology and Geophysics*, 41, 1331–1340 (in Russian).

Hachay, O., E. Novgorodova, et al. (2003a), Resolution research of planshet electromagnetic method for active mapping and heterogeneous geoelectrical media monitoring. *FizikaZemli*, 1, 27–37 (in Russian).

Hachay, O., E. Novgorodova, et al. (2003b), A new method of revealing disintegration zones in the near-hole space of rock massif. *Mining Information Analytical Bulletin, MSSU, Moscow*, 11, 26–29 (in Russian).

Hachay, O., N. Vloch, et al. (2001), 3-D electromagnetic monitoring of rock massif state. *FizikaZemli*, 2, 85–92 (in Russian).

Khachay, A. (1998), System of 3-D seismic and electromagnetic data processing for permafrost geophysical research. *Proc. 7th Int. Conf. Permafrost,* Yellowknife, N.W.T., Canada, 23–27 June 1998, Abstract, 117.

Mazurov, V. (1990), Method of committees in the problems of optimization and classification. Moscow, Nauka. Main redaction of physical and mathematical literature (in Russian).

Nikolis, G. (1989), *Dynamics of Hierarchical Systems*. Mir, Moscow, (in Russian).

Nikolis, G., and I. Prigogine (1979), *Self-organization in Non-equilibrium Systems*. Mir, Moscow (in Russian).

Nikolis, G., and I. Prigogine (1990), *Knowledge of Complexity*. Mir, Moscow (in Russian).

Panin, E., V. Lichachev, et al. (1985), *Structural Deformation Levels of Solid Bodies*. Novosibirsk, SB RAS, Nauka (in Russian).

Samarsky, A., and S. Kurdjumov, (1989), *Paradoxes of Multivariant Nonlinear World Around Us: Hypothesis and Forecasting*. Znanie, Moscow, pp. 8–29 (in Russian).

Shkuratnik, V., and A. Lavrov (1997), *Effects of Memory in Rocks: Physical Regularities, Theoretical Models*. Academy of Mining, Moscow (in Russian).

Vloch, N. (1994), *Management of the Rock Pressure on Mines*. Nedra, Moscow (in Russian).

Vloch, N. (1995), Geomechanical method of effective and secure outworking of ore deposits in mines. *Mining Bulletin*, 4, 28–31 (in Russian).

Vloch, N. (1997), The problem of defining massif state stress of rocks. *Problems of Rock Mechanics*. St. Petersburg, pp. 93–102 (in Russian).

Vloch, N., Y. Lipin, et al. (1994), Choice of the system of outworking and defining its parameters by outworking magnezite using underground methods. *Mining Journal*, 5, 22–25 (in Russian).

8

8. REFLECTION OF PROCESSES OF NONEQUILIBRIUM TWO-PHASE FILTRATION IN OIL-SATURATED HIERARCHIC MEDIUM BY DATA OF ACTIVE WAVE GEOPHYSICAL MONITORING

Olga Hachay[1] and Andrey Khachay[2]

Abstract

A comparison is provided of nonequilibrium effects of independent hydrodynamic and electromagnetic induction on an oil layer and the medium it surrounds. It is known that by some cycles of influence (drainage-steep-drainage), the hysteresis effect on curves of the relative phase permeability in dependence on the porous medium's water saturation is observed. Using the earlier developed 3-D method of induction electromagnetic frequency geometric monitoring, we showed the possibility of defining the physical and structural features of a hierarchic oil layer structure and estimating the water saturation from crack inclusions. This effect allows management of the process of drainage and steeping the oil out of the layer by water displacement. An algorithm was constructed for 2-D modeling of sound diffraction on a porous fluid-saturated intrusion of a hierarchic structure located in layer number J of an N-layered elastic medium.

[1] Institute of Geophysics Ural Branch of Russian Academy of Sciences, Ekaterinburg, Russia
[2] Ural Federal University, Ekaterinburg, Russia

Oil and Gas Exploration: Methods and Application, Monograph Number 72,
First Edition. Edited by Said Gaci and Olga Hachay.
© 2017 American Geophysical Union. Published 2017 by John Wiley & Sons, Inc.

8.1. Introduction

The processes of oil deposit development are linked with the movement of multiphase, multicomponent media, which are characterized by nonequilibrium and nonlinear rheological features. The real behavior of layered systems is defined by the complexity of the rheology of moving fluids and the morphology structure of the porous medium, and also by the great variety of interactions between the fluid and the porous medium [*Hasanov and Bulgakova*, 2003]. It is necessary to take into account these features in order to informatively describe the filtration processes due to the nonlinearity, nonequilibrium, and heterogeneity that are features of real systems. In this way, new synergetic events can be revealed (namely, a loss of stability when oscillations occur, and the formation of ordered structures). This allows us to suggest new methods for the control and management of complicated natural systems that are constructed on account of these phenomena. Thus, the layered system, from which it is necessary to extract the oil, is a complicated dynamical hierarchical system.

8.2. Development of a Mathematical Model Using the Results of Active and Passive Geophysical Monitoring

To construct the mathematical model of a real object, as a priori information it is necessary to use data from active and passive monitoring, which we can obtain during exploitation of the object. The solution of inverse problems has great significance for the oil industry, because the oil layer refers to a number of natural systems that cannot be observed as a whole by direct measurements. The results of research done in 2004 [*Hachay*, 2004] showed that in the evolution of dynamic systems, nonstabilities and their origin play a role in the theory of self-organization or synergetic studies. Information about their manifestation in the oil reservoir from its extraction can only be obtained using monitoring data, which is sensitive to its hierarchic structure. It should be noted that, to study the thin structure of the discrete hierarchic media, geophysical fields are more sensitive, depending on spatial, time, or frequency parameters, namely, electromagnetic and seismic fields. In addition, these fields, excited by local sources due to the geometry of the normal field, have a focusing or localization property that allows the given resolution to be distinguished.

In the Institute of Geophysics, a planshet method of electromagnetic induction research was developed in a frequency-geometrical variant; it differs from the tomography methods and is widely used for mapping and monitoring highly complicated nonstationary geological media for surface and underground (mine) variants. The adaptation of this method to underground research in mining holes to define rock massif structures, their state, and their dynamics according to manmade influences has allowed volume geophysical research in the geological medium to be performed [*Hachay et al.*, 2001; *Hachay et al.*, 2003]. A new complex volume method of electromagnetic induction and seismic (in the dynamic variant) research allows the construction of a volume geoelectrical and elastic model of the rock massif structure. Using this method in mining conditions for deposits of different material content revealed zones of rock massif heterogeneities. The obtained criteria allowed the grading of these zones into zones of hidden fracture and contact (between different modules) zones, which were confirmed by geological and geomechanical data [*Hachay*, 2004, 2007; *Hachay and Khachay*, 2008], and the staged detection of these zones by seismic and electromagnetic data was researched.

In the presented sections (Figure 8.1 a, b), we can see that even during a short period of time (1 week), the most significant change of location of the heterogeneity zones occurs under the influence of explosions.

Analysis of the results of electromagnetic induction monitoring in natural conditions allows the following conclusions to be formulated: the rock massif structure of different material content corresponds to the model of the hierarchic structure of the discrete medium; we could use our system of observation to deduce two hierarchic levels. The disintegration zones, revealed by the electromagnetic monitoring data in the surrounding hole space, are located nonsymmetrically in the roof and in the bottom and are discrete: that is, there are intervals in which the maximum change in the massif, which occurs directly under man-made influence in the morphology of the spatial location of these zones, depending on time, is absent as a whole.

To consider the behavior of the two-phase rock massif in the frame of the model of a hierarchical medium of arbitrary rank, we developed an algorithm for solution of the direct 2-D problem for the seismic field in the dynamic variant. In this way, the model of the local hierarchical heterogeneity of the Lth rank is presented as a porous fluid-saturated inclusion. The hierarchical inclusions of other ranks are presented as elastic heterogeneities in the frame of an approximation, when the parameter Lame $\mu = 0$, either in the inclusions or in the imbedded medium. For that case, the seismic dynamic problem can be considered independently for the cases of the distribution of longitudinal and transversal waves. Here we will consider the first case for the suggested model. The obtained results can be used to determine the joining criteria of seismic research methods for highly complicated media.

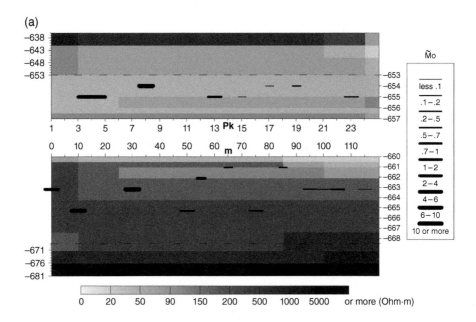

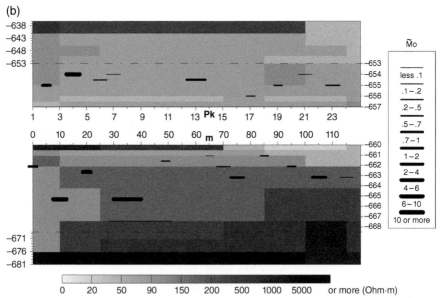

Figure 8.1 Geoelectrical sections for the roof (top) and the ground (bottom) of the 4th ort, horizon –210, Tashtagol mine 2002. a. and b.: two cycles of observations corresponding to 14th and 21st of August 2002, respectively. Frequency 5 kHz. $\tilde{M}_0 = M_0 \times L_0 \times 10^3$, M_0 is the coefficient on which is multiplied the moment of the electrical current line, which is equivalent in the field to the influence of the zone of geoelectrical heterogeneity and is proportional to the ratio of conductivity difference in the embedded medium; L_0 is the length of the current line; resistivity of the embedded section is presented in om.m. On the vertical axes, values presented are in m (absolute marks), on the horizontal axes the length of the hole in pickets (Pk) and in meters is presented.

8.3. Algorithm of Modeling for Longitudinal wave Propagation in the Medium with Hierarchic Inclusions

A concept was suggested in a paper by *Hachay and Khachay* [2013] for solution of the direct problem for the 2-D case of longitudinal wave propagation through a local elastic heterogeneity with a hierarchical structure, located in the *J*th layer of an *N*-layered medium. Let us extend this to the case, on the *L*th hierarchical level, of the occurrence of a porous fluid-saturated inclusion.

$$
\frac{\left(k_{1jil}^2 - k_{1j}^2\right)}{2\pi} \iint\limits_{S_{Cl}} \varphi_l(M) G_{Sp,j}\left(M, M^0\right) d\tau_M + \frac{\sigma_{ja}}{\sigma_{jil}} \varphi_{l-1}^0\left(M^0\right) -
$$

$$
-\frac{\left(\sigma_{ja} - \sigma_{jil}\right)}{\sigma_{jil} 2\pi} \oint\limits_{Cl} G_{Sp,j} \frac{\partial \varphi_l}{\partial n} dc = \varphi_l\left(M^0\right), \ M^0 \in S_{Cl}
$$

$$
\frac{\sigma_{jil}\left(k_{1jil}^2 - k_{1j}^2\right)}{\sigma\left(M^0\right)2\pi} \iint\limits_{S_{Cl}} \varphi_l(M) G_{Sp,j}\left(M, M^0\right) d\tau_M + \varphi_{l-1}^0\left(M^0\right) -
$$

$$
-\frac{\left(\sigma_{ja} - \sigma_{jil}\right)}{\sigma\left(M^0\right)2\pi} \oint\limits_{Cl} G_{Sp,j} \frac{\partial \varphi_l}{\partial n} dc = \varphi_l\left(M^0\right), \ M^0 \notin S_{Cl},
$$

(8.1)

where $G_{Sp,i}(M, M^0)$ is the source function of seismic field, which coincides with the function from the paper by *Hachay and Khachay* [2013]. $k_{1jil}^2 = \omega^2\left(\sigma_{jil}/\lambda_{jil}\right)$ is the wave number for the longitudinal wave. In the equations, the index *ji* indicates the features within the heterogeneity membership, *ja* is out of the heterogeneity, $l = 1...L\text{-}1$ is the number of the hierarchic level, $\vec{u}_l = grad\varphi_l$, φ_l^0 is the potential of the normal seismic field in the layered medium, when the heterogeneity of the previous rank is absent. If $l = 2...L$ $\varphi_l^0 = \varphi_{l-1}$, if $l = 1$, $\varphi_l^0 = \varphi^0$, this coincides with the expression from the paper by *Hachay and Khachay* [2013].

If, with transition to the next hierarchic level, the axis of 2-D does not change and only the geometries of the embedded structures change, then likewise [*Hachay and Khachay*, 2013] we can develop the iteration process for modeling of the seismic field (the case of only a longitudinal wave forming). The iteration process is related with the modeling of the displacement vector by transition from the previous hierarchic level to the next level. Within each hierarchic level, the integral-differential equation and the integral-differential representation can be evaluated by algorithm (1). If on a certain hierarchic level, the structure of the local heterogeneity is divided into a number of heterogeneities, the integrals in the expressions (1) are evaluated on all those heterogeneities. If $l=L$, then within the heterogeneities of the previous hierarchical level there occurs porous fluid-saturated heterogeneity. In that case, the system (1), taking account of *Frenkel* [1944], can be rewritten as follows:

$$\frac{\left(k_{1jil}^2 - k_{1j}^2\right)}{2\pi} \iint\limits_{S_{ol}} \varphi_l(M) G_{Sp,j}\left(M, M^0\right) d\tau + \frac{\sigma_{ja}}{\sigma_{jil}} \varphi_{l-1}^0\left(M^0\right) -$$

$$\frac{\left(\sigma_{ja} - \sigma_{jil}\right)}{\sigma_{jil} 2\pi} \oint\limits_{C_l} G_{Sp,j} \frac{\partial \varphi_l}{\partial n} dc = \left(\varphi_l\left(M^0\right) + \alpha\, p_2\right), M^0 \in S_{0l}, l = L$$

$$\frac{\sigma_{jil}\left(k_{1jil}^2 - k_{1j}^2\right)}{2\pi} \iint\limits_{S_{ol}} \varphi_l(M) G_{Sp,j}\left(M, M^0\right) d\tau + \frac{\sigma_{ja}}{\sigma_{jil}} \varphi_{l-1}^0\left(M^0\right) -$$

$$\frac{\left(\sigma_{ja} - \sigma_{jil}\right)}{\sigma\left(M^0\right) 2\pi} \oint\limits_{C_l} G_{Sp,j} \frac{\partial \varphi_l}{\partial n} dc = \varphi_l\left(M^0\right), M^0 \in S_{0l}, l = L,$$

(8.2)

where $\alpha = 1 - \chi - \dfrac{K}{K_0}$, $K = \lambda$ is modulus of uniform compression, χ is porosity, K_0 is true modulus of phase compression, pore hydrostatic pressure p_2.

If $l = L + 1$ and on the next level the heterogeneity is again elastic, then for further continuation of the iteration process we can again use the system (1).

8.4. Conclusions

The algorithm developed for modeling, and the method of mapping and monitoring a heterogenic highly complicated two-phase medium can be used for managing viscous oil extraction in mining conditions and light oil in subhorizontal boreholes. The demand for effective economic parameters and fuller extraction of oil and gas from deposits dictates the necessity of developing new geotechnology based on the fundamental achievements in the area of geophysics and geomechanics [*Oparin et al.*, 2010].

References

Frenkel, Y. (1944), On the theory of seismic and seismo-electrical phenomena in a humid soil. *Izvestiya AN USSR, Geographic and Geophysical Series*, 8(4), 133–150 (in Russian).
Hachay, O. (2004), The phenomenon of self-organization in rock massifs by man-made influence. *Physical Mesomechanics*, 33/34, 32–37 (in Russian).

Hachay, O. (2007), Geophysical monitoring of the rock massif state using the paradigm of physical mesomechanics. *FizikaZemli*, 4, 58–64 (in Russian).

Hachay, O., and Khachay, A. (2013), Modeling of electromagnetic and seismic field in hierarchic heterogeneous media. *Bulletin YuURGU, Computational Mathematics and Informatics*, 2(2), 48–55 (in Russian).

Hachay, O., and Khachay, O. (2008), Theoretical approaches to justification of the system of geophysical control of the geological medium state by man-made influence. *Mining Information-Analytical Bulletin*, 1, 161–169 (in Russian).

Hachay, O., E. Novgorodova, and O. Khachay (2003), New method of revealing disintegration zones in near-hole space of rock massifs of different material content. *Mining Information-Analytical Bulletin*, 11, 85–92 (in Russian).

Hachay, O., N. Vloch, et al. (2001), 3-D electromagnetic monitoring of massif rocks state. *Fizika Zemli*, 2, 85–92 (in Russian).

Hasanov, M., and G. Bulgakova (2003), Nonlinear and nonequilibrium effects in rheological complicated media. Institute of Computer Research, Moscow, Izhevsk (in Russian).

Oparin, V., et al. (2010), *Geomechanical and Technical Basement of Increasing of Oil Layers Recovery by Oscillation-Wave Technologies*, Nauka, Novosibirsk (in Russian).

9. DEFINING THE SURFACE OF THE FLUID-SATURATED POROUS INCLUSION IN THE HIERARCHIC LAYERED-BLOCK MEDIUM ACCORDING TO ELECTROMAGNETIC MONITORING DATA

Olga Hachay[1] and Andrey Khachay[2]

Abstract

Definition of the fluid-saturated porous inclusion in the hierarchic layered-block medium is linked with a problem of constructing approaches for the solution of inverse problems. For that inverse solution, electromagnetic monitoring data that are sensitive to a geological model with inclusions of hierarchical structure are used. A three-stage approach is suggested for the interpretation of electromagnetic data, which is widely used for 3-D interpretation of mapping in the frame of the frequency-distance active electromagnetic method. Here, we have written new integral-differential equations for the third stage of interpretation, named the theoretical inverse problem solution for 2-D electromagnetic field in the frame of an n-layered model with a hierarchic inclusion of the k rank.

9.1. Introduction

The results of last year's research [*Panin et al.*, 1985] showed that in the evolution of dynamic systems of nonstabilities, study of the theory of self-organization or synergetic theory plays the main role. Information on the manifestation of

[1] *Institute of Geophysics Ural Branch of Russian Academy of Sciences, Ekaterinburg, Russia*
[2] *Ural Federal University, Ekaterinburg, Russia*

Oil and Gas Exploration: Methods and Application, Monograph Number 72,
First Edition. Edited by Said Gaci and Olga Hachay.

nonstabilities in the oil reservoir derived from it developing can be obtained only by using monitoring data that is sensitive to its hierarchic structure. We must note that for more sensitive and higher resolution research of the detailed structure of the discrete hierarchic media, we can consider the geophysical fields, which depend either on spatial coordinates, or on time or frequency, namely the seismic and electromagnetic fields. In addition, these fields, when excited by controlled sources, due to the geometry of the normal field, possess a focus and localization feature that allows the given resolution to be achieved [*Hasanov and Bulgakova*, 2003].

By studying space-time structural change, and the physical features of the geological medium or rock massif and the associated stress-deformed and phase state, we observed that the model of a layered block with inclusions is in fact more complicated; it presents a two-rank chain in the common hierarchic heterogeneous medium model. The model of a hierarchic heterogeneous medium describing the deformation processes and destruction inside a geophysical medium was first suggested by academician *M.A. Sadovskiy* [1987]. A set of papers from scientists of the Institute of Earth's Physics of the Russian Academy of Sciences were devoted to the development and use of the hierarchic block on a qualitative level [*Rodionov et al.*, 1989]. In understanding the formation and development of the hierarchical structural deformation levels in solid bodies, a significant role was played by the theoretical and experimental results obtained from specimens [*Panin et al.*, 1985]. This approach [*Nikolis and Prigozhin*, 1979] was based on the assumptions of dissipative structures in nonequilibrium systems, using these results. *N. A. Karaev* [*Karaev and Rabinovitch*, 2000; *Karaev*, 2000] summarized the results of seismic research devoted to mapping the areas of the Earth's crust with the heterogenic type structure. Heterogeneity, according to the author, is the main peculiarity of rocks, which is stipulated by the irregularity of the spatial distribution of geological heterogeneities as inclusions on all scales; that is, essentially the study of the structure and dynamics of the heterogenic areas of the Earth's crust must be based upon the use of hierarchic model presentations. The phenomenon of disintegration zones around underground holes, linked with the discreteness and medium fragmentation, was described by academician E.I. Shemjakin and his coauthors [1986] and was accorded the status of a discovery [*Shemjakin et al.*, 1992]. These phenomena are presented in both shallow mines (up to 500 m) and in deep mines (more than 500 m). The results of geological-geophysical research on super-deep boreholes show that with increasing depth the complexity of the geological structure does not decrease. Finally, the problem of monitoring and forecasting the state of the geological medium demands a very accurate choice of methods of presenting 3-D medium research that allows space-time (frequency) scaling and focusing. In the paper by *Hachay and Khachay* [2013] an algorithm of 3-D modeling of the electromagnetic field was developed for an arbitrary type of exciting source of an N-layered medium with hierarchic conductive inclusion, located in the jth layer.

To construct a mathematical model of a real object, as a priori information, it is necessary to use data from active and passive monitoring, obtained during the exploitation of the object. The solution of inverse problems is of great significance for the oil industry, because the reservoir refers to a number of natural systems that cannot be observed directly or as a whole [*Hasanov and Bulgakova*, 2003].

9.2. Inverse Problem Solution Algorithm of 2-D Electromagnetic Monitoring Data from the Layered-Block Medium with Hierarchic Structure

The paper by *Hachay* [1994] suggested a conception of a staged interpretation of alternating electromagnetic field. In the first stage, the parameters of the normal section, or the parameters of the 1-D nonmagnetic medium that contains anomaly conductive or magnetic inclusions are defined. In the second stage, fitting of the anomalous alternating electromagnetic field is performed by a system of singular sources, located in the horizontal-layered medium with geoelectrical parameters defined in the first stage. In the third stage, the theoretical inverse problem is solved; that is, for the given geoelectrical parameters of the containing medium, for the set of heterogeneity parameters the contours of the heterogeneity are defined. For this, we obtained the explicit integral-differential equations of the theoretical inverse problem of the distribution of 2-D and 3-D alternating and 3-D stationary electromagnetic fields in the frame of models of a conductive or magnetic body located in the νth layer of a conductive n-layered half-space.

Here, using the approach from the papers by *Hachay* [1989, 1990], we will develop an algorithm for deriving the equation of the theoretical inverse problem for alternating electromagnetic field (scalar case) for the model of a conductive hierarchic heterogeneity of the kth rank, located in the νth layer of a conductive n-layered half space, together with its realization as an iteration algorithm.

Let the simply connected area D from Euclid space R^2, bounded by continuous differentiated closed curve ∂D, be located in the νth layer of the n-layered half-space. Let this area contain in it K noncoaxial simply connected hierarchic inclusions, bounded by continuous differentiated closed curves ∂D_k and spread parallel to the axis OX. The boundaries l_j of the layers Π_j ($j=1,\ldots, n$) are parallel to the axis OY of the plane XOY of the Cartesian coordinate system. The axis OZ is directed vertically down. We will locate the coordinate origin on the upper boundary of the surface of the first layer and combine it with the point that is a projection on the axis OY of the point to which relatively the domain D is star-like. Let $U(y, z)$-complex be twice the continuous differentiated function that satisfies relatively the 2-D scalar Helmholtz equation:

$$\Delta U + c(M)U = -f(M), \tag{9.1}$$

where $\Delta = \dfrac{\partial^2}{\partial y^2} + \dfrac{\partial^2}{\partial z^2}$;

$$c(M) = \begin{cases} c_j ; M \in \Pi_j \setminus \bar{D}(j = 0,...n) \\ c_{ak}; M \in D_k \ (k = 1,...K) \end{cases}. \tag{9.2}$$

Let the function $U^1(y, z)$ satisfy the equation

$$\Delta U^1 + p(M)U^1 = -f(M) \tag{9.3}$$

$$p(M) = \begin{cases} c_j ; M \in \Pi_j \setminus \bar{D}(j = 0, ...n) \\ c_v; M \in D_k \ (k = 1, ...K) \end{cases}. \tag{9.4}$$

We will first consider the case $k = 1$. For $M \in R^2 \setminus \bar{D}(j = 0, ...n)$ we will define

$$U^+(M) = U(M) - U^1(M). \tag{9.5}$$

The function $U^+(M)$ satisfies equation (1). On the boundaries l_j of the layers Π_j the boundary conditions are fulfilled as follows:

$$\begin{aligned} U_j &= U_{j+1}; \\ U_j^+ &= U_{j+1}^+; \ M \in l_j \ (j = 1, ..., n-1); \\ U_j^1 &= U_{j+1}^1; \end{aligned} \tag{9.6}$$

$$\begin{aligned} b_j \frac{\partial U_j}{\partial n} &= b_{j+1} \frac{\partial U_{j+1}}{\partial n}; \ b_j \frac{\partial U_j^+}{\partial n} = b_{j+1} \frac{\partial U_{j+1}^+}{\partial n}; \\ b_j \frac{\partial U_j^1}{\partial n} &= b_{j+1} \frac{\partial U_{j+1}^1}{\partial n}; \ M \in l_j \end{aligned} \tag{9.7}$$

where b_j is the complex coefficients $(j = 0, ..., n)$ and for the common case $b_j \neq b_{j+1}$; on the contour ∂D_k by

$$k = 1; \ U_v = U_v^+ + U_v^1. \tag{9.8}$$

The function U_v satisfies the equation

$$\Delta U_v + c_v(M)U_v = -f(M). \tag{9.9}$$

U_v^+ is function U^+ in the layer $\Pi_v \notin D$; U_v^1 is a function U^1 in the layer $\Pi_v \notin D$; in the domain D by $k = 1$

$$U_a = U_a^+ + U_a^1; M \in \bar{D}; \Delta U_a + c_a U_a = 0. \tag{9.10}$$

Boundary conditions on $\partial D(k=1)$:

$$U_a^+ = U_v^+ b_a \frac{\partial U_a}{\partial n} - b_v \left(\frac{\partial U_v^+}{\partial n} + \frac{\partial U_v^1}{\partial n} \right) = 0. \tag{9.11}$$

By $M \to \infty$ the functions $U(M), U^+(M), U^1(M)$ satisfy the radiation condition [*Stratton*, 1948]. The algorithm of calculation for the function U^1 is written in *Hachay* [1994]. Let us introduce the function $G(M, M_0)$, which satisfies the equation

$$\Delta G + p(M)G = -\delta(M, M_0) \tag{9.12}$$

and the boundary conditions (6, 7) by $M \to \infty$ the function G satisfies the radiation condition [*Stratton*, 1948], by $M \to M_0$ the function G has a singularity as a type $\ln 1/\rho(M, M_0)$:

$$\rho(M, M_0) = \sqrt{(y - y_0)^2 + (z - z_0)^2}. \tag{9.13}$$

The algorithm of calculation for function G for the case when the domain D is located in the v-th layer is illustrated in *Hachay* [1994]. Let us introduce the function G^a, which coincides with the fundamental solution of equation (2) by $k=1$. Let us use Green's formula [*Stratton*, 1948] for the pair functions $U^+, G; \left(M \in R^2 \backslash \bar{D}, M_0 \in \Pi_i \right)$ in each layer $\Pi_j (j = 0, ..., n)$. Let us provide the procedure similarly [*Dmitriev*, 1965]: let us multiply the obtained expressions for each layer with b_j reciprocally, $j = 0, ..., n$ and fold term-wise taking account of equations (1)–(4), (6), and (7).

As a result, we obtain

$$2\pi U^+(M_0) = -\left(\frac{b_v}{b_i} \right) \int_{\partial D} \left(U_v^+(M) \frac{\partial G(M, M_0)}{\partial n} - G(M, M_0) \frac{\partial U_v^+}{\partial n} \right); \tag{9.14}$$

$$M \in \Pi_v; M_0 \in \Pi_i;$$

In the domain D let us use Green's formula for the pair functions $U_a(M), G^a(M, M_0)$ (10). As the result, we obtain

$$0 = \int_{\partial D} \left(U_a(M) \frac{\partial G^a(M, M_0)}{\partial n} - G^a(M, M_0) \frac{\partial U_a}{\partial n} \right) dl. \tag{9.15}$$

Let us summarize expressions (14) and (15), taking into account (10)–(11), and also the expression [*Hachay*, 1994]

$$0 = \left(-\frac{b_v}{b_i} \int_{\partial D} (U_v^1(M) \frac{\partial G(M, M_0)}{\partial n} - G(M, M_0) \frac{\partial U_v^1}{\partial n} \right) dl; \tag{9.16}$$

$$M \in \bar{D}; M_0 \in \Pi_i;$$

Then we obtain

$$
\begin{aligned}
2\pi U^+\left(M_0\right)=\int_{\partial D}\Bigg(\left(U_v^+\left(M\right)+U_v^1\left(M\right)\right)\Bigg(\frac{\partial G^a\left(M,M_0\right)}{\partial n}\\
-\left(\frac{b_v}{b_i}\right)\frac{\partial G\left(M,M_0\right)}{\partial n}\Bigg)-b_v\left(\frac{\partial U_v^+}{\partial n}+\frac{\partial U_v^1}{\partial n}\right)\left(\left(\frac{1}{b_a}\right)G^a\left(M,M_0\right)\right.\\
-\left(\frac{1}{b_i}\right)G\left(M,M_0\right)\Bigg)\Bigg)dl.
\end{aligned}
$$

$$(9.17)$$

Equation (17) is an explicit equation of the theoretical inverse problem for the 2-D scalar Helmholtz equation in the frame of the model of the layered medium with a homogeneous inclusion for the given values of the boundary conditions [*Hachay*, 1994, 1989, 1990]. As a result of solving the integral-differential equation (17) concerning the function $r(\varphi)$, which describes the contour of the homogeneous object we seek, we can define it by the known values of the physical parameters of the containing medium and the desired object and also for the given values of functions $U^+, G, G^a, U_v^+, U_v^1$.

Let us consider the case of E-polarization. The coefficients are

$$
b_v = b_i = b_a = 1. \tag{9.18}
$$

Let us replace in (17) $U \to E_x; U^+ \to E_x^+; U^1 \to E_x^1$, where $E_x(y,z,\omega)$ is a component of the electromagnetic field, directed along the spreading of the body, $G_E = G, G_E^a = G^a$ are functions that are defined from the solution of the boundary problems, written above by satisfying the condition (18) [*Hachay*, 1994]. From this, in (2):

$$
c(M) = k^2(M) = i\omega\mu_0\sigma(M), \tag{9.19}
$$

where i is image unit, ω is circled frequency, $\mu_0 = 4\pi \cdot 10^{-7}\ Hn/m$, $\sigma(M)$ is conductivity of the medium at the point M; in (2):

$$
c(M) = k^2(M) = \begin{cases} k_j^2 = i\omega\mu_0\sigma_j;\ M \in \Pi_j \setminus \overline{D}(j = 0, ..., n); \\ k_a^2 = i\omega\mu_0\sigma_a;\ M \in \overline{D} \end{cases} \tag{9.20}
$$

In expression (4)

$$
p(M) = k^2(M) = \begin{cases} k_j^2 = i\omega\mu_0\sigma_j;\ M \in \Pi_j \setminus \overline{D}(j = 0, ..., n); \\ k_v^2 = i\omega\mu_0\sigma_v;\ M \in \overline{D} \end{cases} \tag{9.21}
$$

$$2\pi E_x^+ \left(M_0 \right) = \int_{\partial D} \left(\left(E_{xv}^+ \left(M \right) + E_{xv}^1 \left(M \right) \right) \left(\frac{\partial G_E^a \left(M, M_0 \right)}{\partial n} \right. \right.$$

$$\left. - \frac{\partial G_E \left(M, M_0 \right)}{\partial n} \right) - \left(\frac{\partial E_{xv}^+}{\partial n} + \frac{\partial E_{xv}^1}{\partial n} \right) \left(G_E^a \left(M, M_0 \right) \right. \tag{9.22'}$$

$$\left. \left. - G_E \left(M, M_0 \right) \right) \right) dl.$$

Let $k = 2$, such that the object we are seeking has a hierarchic structure: ∂D_1 is a contour of the outer inclusion with conductivity σ_1 and the contour of the inner inclusion ∂D_2 with conductivity σ_2. The inclusions are not coaxial. It is necessary to construct the two contours. This problem can be solved in two stages. The first stage: for solution of that problem for the case of E-polarization in the expression (20) $\sigma_a = \sigma_1$, in the equation (22): $\partial D = \partial D_1; dl = dl_1; G_E^a = G_E^{a1}; \frac{\partial}{\partial n} \left(G_E^a \right) = \frac{\partial}{\partial n} \left(G_E^{a1} \right)$.

Solving equation (22) concerning the function $r_1(\varphi)$, which describes the contour ∂D_1, we calculate the functions

$$U \to E_x; U^+ \to E_x^+; U^1 \to E_x^0 \tag{9.23}$$

using the algorithm of the direct problem [*Dmitriev*, 1965] inside and outside the heterogeneity, embedded in the layered medium, where E_x^0 is the electrical component of the electromagnetic field in the layered medium without the heterogeneity.

$$U(M) = U^1(M) + \frac{k_v^2 - k_{a1}^2}{2\pi} \iint_{S1} U(P) G_E \left(M, P \right) dS_1; \; M \in S_1 \tag{9.24}$$

$$U(M_i) = U^1(M_i) + \frac{k_i^2 - k_{a1}^2}{2\pi} \iint_{S1} U(P) G_E \left(M_i, P \right) dS_1; \; M_i \in \Pi_i \tag{9.25}$$

Thus, the first iteration cycle is finished and we begin the second iteration cycle $k = 2$:

$$E_x^{1(k-1)} \left(M_i \right) \to U \left(M_i \right); \tag{9.26}$$

Corresponding to the calculated function $U(M_i)$ (25), in equation (20): $\sigma_a = \sigma_2$, in equation (22):

$$\partial D = \partial D_2; G_E^a = G_E^{a2}; \frac{\partial}{\partial n} \left(G_E^a \right) = \frac{\partial}{\partial n} \left(G_E^{a2} \right), \; dl = dl_2. \tag{9.26'}$$

Equation (22) can be rewritten as follows:

$$2\pi E_x^+(M_0) = \int_{\partial D} \left(\left(E_{xv}^+(M) + E_{xv}^{1(k-1)}(M) \right) \left(\frac{\partial G_E^a(M,M_0)}{\partial n} \right. \right.$$

$$\left. - \frac{\partial G_E(M,M_0)}{\partial n} \right) - \left(\frac{\partial E_{xv}^+}{\partial n} + \frac{\partial E_{xv}^{1(k-1)}}{\partial n} \right) \left(G_E^a(M,M_0) \right. \tag{9.27}$$

$$\left. \left. - G_E(M,M_0) \right) \right) dl.$$

We will solve equation (27) concerning the function $r_2(\varphi)$ that describes the contour ∂D_2. If $k=2$, then the problem is solved, if $k>2$, $k=k+1$ and the iteration process continues. We calculate the functions

$$U^{k-1} \to E_x^{k-1}; U^{+(k-1)} \to E_x^{+(k-1)}; U^{1(k-2)} \to E_x^{1(k-2)} \tag{9.28}$$

using the algorithm for solution of the direct problem inside and outside the hierarchic heterogeneity of the rank $k-1$, located in the layered medium (the physical parameters of the layered medium are the same).

$$U^{k-1}(M) = U^{1(k-2)}(M) + \frac{k_v^2 - k_{a(k-1)}^2}{2\pi} \iint_{S(k-1)} U^{k-1}(P) G_E(M, P) dS_{(k-1)}; \tag{9.29}$$

$$M \in S_{(k-1)}$$

$$U^{k-1}(M_i) = U^{1(k-2)}(M_i) + \frac{k_i^2 - k_{a(k-1)}^2}{2\pi} \iint_{S(k-1)} U^{(k-1)}(P) G_E(M_i, P) dS_{(k-1)}; \tag{9.30}$$

$$M_i \in \Pi_i$$

In expression (20): $\sigma_a = \sigma_k$, in expression (27):

$$\partial D = \partial D_k; G_E^a = G_E^{ak}; \frac{\partial}{\partial n}\left(G_E^a \right) = \frac{\partial}{\partial n}\left(G_E^{ak} \right); dl = dl_k. \tag{9.30'}$$

If we replace the calculated function $U^{k-1}(M_i)$ (30) with $E_x^{1(k-1)}$, then we solve equation (27), concerning the function $r_{(k-1)}(\varphi)$, which describes the contour $\partial D_{(k-1)}$, $k=k+1$. The iteration process (27)–(30') continues up to $k=K$.

9.3. Conclusions

The definition of the fluid-saturated porous inclusion in the hierarchic layered-block medium is linked with a problem of constructing approaches for solution of the inverse problems. Here, we developed an algorithm of solution using electromagnetic monitoring data in the frame of the geological medium model with hierarchic inclusions. Earlier, a three-stage approach was suggested for the interpretation of electromagnetic data, which is widely used for 3-D interpretation of mapping in the frame of the frequency-distance active electromagnetic method. Here we have written a new integral-differential equation for the third stage of interpretation, named the theoretical inverse problem solution for 2-D electromagnetic field in the frame of an n-layered model with a hierarchic inclusion of the k rank. The solution of inverse problems is of great significance for the oil industry, because the oil layer refers to a number of natural systems that cannot observed directly or as a whole.

References

Dmitriev, V. (1965), Diffraction of plane electromagnetic field on cylindrical bodies located in layered media. *Computational Methods and Programming*. MSU, Moscow, 3, 307–315 (in Russian).

Hachay, O. (1989), On the interpretation of 2-D alternating and 3-D stationary anomalies of electromagnetic field. *Izvestija AN USSR, Fizika Zemli*, 10, 50–58 (in Russian).

Hachay, O. (1990), On the inverse problem solution of 3-D alternating electromagnetic fields. *Izvestija AN USSR, Fizika Zemli*, 2, 55–59 (in Russian).

Hachay, O. (1994), Mathematical modeling and interpretation of alternating electromagnetic field in the heterogeneous crust and upper mantle of the Earth. *Professorial Dissertation*, IGF UB RAS, Sverdlovsk (in Russian).

Hachay, O., & A. Khachay (2013), Modeling of electromagnetic and seismic field in hierarchically heterogeneous media. *Bulletin of YuURGU, Computational Mathematics and Informatics*, 2(2), 48–55 (in Russian).

Hasanov, M., and G. Bulgakova (2003), *Non-linear and non-equilibrium effects in rheologically complicated media*. Institute of Computer Research, Moscow-Izhevsk (in Russian).

Karaev, N. (2000), *Seismic heterogeneity of the Earth's crust and problems of the interpretation in the close zone of the results of regional observations*. In *Non-classic Geophysics*, Saratov, pp. 30–32 (in Russian).

Karaev, N., and G. Rabinovitch (2000), *Ore Seismicity*. Geoinformmark, Moscow (in Russian).

Nikolis, G., and I. Prigozhin (1979), Self-organization in non-equilibrium systems. Mir, Moscow (in Russian).

Panin, V., et al. (1985), *Structural Deformation Levels of Solid Bodies*. Nauka, Novosibirsk, (in Russian).

Rodionov, V., et al. (1989), On the modeling of natural objects in geomechanics. In *Discrete Features of the Geophysical Medium*. Nauka, Moscow, pp. 14–18 (in Russian).

Sadovskiy, M. et al. (1987), *Deformation of Geophysical Medium and Seismic Process*. Nauka, Moscow (in Russian).

Shemjakin, E., et al. (1986), Effect of zone disintegration of rocks around the underground holes. *DAN USSR*, 289(5), 830–832 (in Russian).

Shemjakin, E., et al. (1992), Effect of zone disintegration of rocks around underground holes. *Discovery No. 400, Bulletin of Inventions*, 1 (in Russian).

Stratton, D. (1948), *Theory of Electromagnetism*. OGIZ, Moscow-Leningrad (in Russian).

10. HARDWARE AND SOFTWARE SYSTEM FOR RESEARCH IN OIL AND GAS BOREHOLES

Yury G. Astrakhantsev, Nadezhda A. Beloglazova, and Eugenia Bazhenova

Abstract

Geoacoustic signals occur naturally in boreholes because of the dynamic processes located in the volume of near borehole space. The processes are linked with fluid motions (water, oil) or cracks forming inside the rocks. The measurement of these signals in different frequency bands gives the opportunity for defining the acoustic oscillations source. The measurements of three components of the acoustic field give the information of the source location relative to the borehole axis. The basis of the method of three-component measurements of geoacoustic signals on oil and gas deposits is considered. Information is provided about the unique equipment, the method of processing, and qualitative interpretation. The effectiveness of three-component geoacoustic research for control of the hydrocarbon deposit exploitation is shown.

10.1. Introduction

Development of the method of three-component measurements of geoacoustic signals on oil and gas deposits is linked with research on geoacoustic signals inside the volume with sediment rocks. During exploration and exploitation of oil and gas deposits in the reservoir rock, natural geoacoustic signals occur in the enclosing rocks exposed in the boreholes [*Trojanov, Astrakhantsev et al.*, 2000]. These reflect different and complicated processes in the rocks area in the borehole volume. The rocks in the sediment thickness can be considered

Institute of Geophysics Ural Branch of Russian Academy of Sciences (UB RAS), Ekaterinburg, Russia

Oil and Gas Exploration: Methods and Application, Monograph Number 72,
First Edition. Edited by Said Gaci and Olga Hachay.
© 2017 American Geophysical Union. Published 2017 by John Wiley & Sons, Inc.

a system that includes the solid skeleton and filled medium, which is in thermo-dynamic equilibrium. Thus, each component that constitutes the system (solid skeleton, oil, water, and gas) can be the source of acoustic oscillations [*Dobrinin*, 1965; *Dobrinin, Gorodnov et al.*, 2001]. Drilling boreholes and using them for oil and gas extraction by different technological methods leads to the violation of that state, which is accompanied by an increase in the occurrence of new mecha-nisms of the generation of acoustic oscillations, which are registered in the sec-tions opened by the boreholes.

Research on natural geoacoustic signals in boreholes allows us to obtain prin-cipally new information about the geological structure of the Earth's crust, the character of the behavior of the dynamic processes in it, and its activity. Measurement of these signals is a major diagnostic tool for oil and gas borehole research.

On the basis of research on the space-time distribution of amplitude-frequency characteristics of geoacoustic signals at the Institute of Geophysics of the Ural Branch of the Russian Academy of Sciences, a hardware-software system for research into natural geoacoustic signals, called geoacoustic emissions (GAE), has been developed.

10.2. Physical Basis of the Method

Let us consider first the solid rock. It has features such as a discrete hetero-genic structure with pores and cracks, which is in a stressed state (with local over-stresses on the heterogeneities, defects) [*Vorobjev, Tonkonogov, et al.*, 1972]. The specific combination of that structure and the stress distribution leads to a quasi-stable state, when at lithostatic pressures up to 100 MPa and higher the rocks respond to negligible displacement deformations of about 10^{-7} m (earth tides) and 10^{-10} m (the Earth's own oscillations, storm microseisms) with acous-tical activity, which reflects the deformation processes and microdisintegration, the origin of new defects, and cracks in the volume of the geoacoustic medium [*Trojanov*, 2000; *Viktorov*, 1981]. As shown by the results of three-component measurements of geoacoustic emissions in deep and super-deep boreholes, the structures of intense disintegration, cracks, and dynamic activity can be revealed by anomalous large values of GAE amplitudes.

A significant role in the activation of deformation processes and therefore geoacoustic emissions is played by the rock's fluid saturation, which leads to a reduction of the rock strength due to reduction of the inner friction, electrochem-ical processes, and other factors [*Ivanov*, 2010; *Tiab and Donaldson*, 2009].

Drilling of oil and gas areas and exploitation of oil and gas deposits disturb the original scheme of pressures, the fluid layer gas saturation, and the tempera-ture distribution. As a result, the dynamics of the rock's solid skeleton and, consequently, the permeability of reservoirs and the intensity of geoacoustic emissions are characteristics that can carry information on man-made influences [*Trojanov and Beloglazova*, 2003; *Trojanov, Igolkina et al.*, 2012].

Selection of the number of channels and their frequency range is carried out using the results of careful laboratory research in the spectral (amplitude-frequency) signal content. Estimation of the geoacoustic signals' informativity about the processes in oil and gas boreholes, including the character of layer-reservoir saturation, is provided [*Trojanov and Astrakhantsev*, 2009; *Trojanov and Astrakhantsev*, 2005]. For sedimentary rocks the integral level of GAE is limited by high frequencies, up to 0.5 kHz.

A typical spectrum of geoacoustic signals from the analogue recording on the magnetic carrier in the thickness of the sedimentary rocks (water-saturated layer-reservoir) is presented in Figure 10.1 (graph 1). The existence of higher frequencies in the spectrum, greater than 0.5 kHz, is an evidence of another source of acoustic signals. In the range of frequencies 0.5–1.5 kHz, the occurrence of geoacoustic signals may be linked with the fluid motion with the gaseous factor (Figure 10.1, graph 2). For the intervals with gas extraction along the wellbore, the characteristic form of the spectrum is shown in Figure 10.1, graph 3. For the case of intense gas extraction, the GAE spectrum covers the whole frequency range (Figure 10.1, graph 4). In that case, we cannot divide the signal into components because the values of the amplitudes are very large. Thus, using the peculiarities of the distribution of geoacoustic signals in different frequency ranges, it is possible to diagnose the character of reservoir saturation. But this is only a general approach to solving the problem, and for a detailed interpretation of the obtained data, it is necessary to study not only the measured data but also the whole complex of calculated GAE parameters, taking into account geological and technological factors. It must be noted that for each specific problem, we used only those parameters of GAE that are the most informative for the given

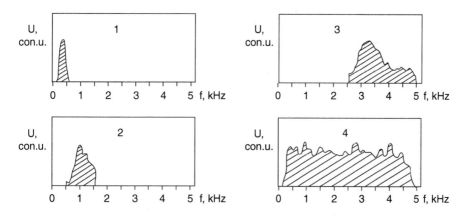

Figure 10.1 Spectra of GAE. 1: water-saturated layer, 2: oil-saturated (fluid with gaseous factor) layer, 3 and 4: existence of gas extraction zones. The vertical axis U is the value of amplitude in conventional units.

case [*Trojanov, Igolkina et al.*, 2012]. For quantitative interpretation, a unique methodical and interpretation system must be developed.

10.3. New Hardware for Geoacoustic Digital Measurements

At the present time, at the Institute of Geophysics UB RAS some modifications to digital devices have been developed together with software for the registration and processing of geoacoustic signals. The latest modification developed at the institute is the BN-4008 device.

This device is used for measuring three components of the vibration vector in the borehole, which is stipulated by natural causes. The vibration of the borehole device may be due to the motion of the water, oil, gas, tectonic disintegrations in the wellbore, and the borehole environment. Measurement of the geoacoustic signals during exploitation of the hole is provided through the casing steel column, which is often magnetized. This leads to the deletion of magnetic labels on the cable and to mistakes being made when defining the depth of the borehole device's location. Therefore, it is necessary to use the gamma-ray block to define the depth of layers with increased natural radioactivity. With increasing borehole depth of the hydrocarbon deposits, the question of the thermal stability and reliability of the borehole device arises, and high noise immunity of the connection channel between the borehole and the surface device is especially necessary. Temperature measurements in the borehole at the moment of measuring the geoacoustic signals allow a more unambiguous interpretation of the data obtained.

Geoacoustic signals stipulated by the medium's vibration can change over a wide amplitude range in different objects. To arrange a good resolution of the device, it is necessary to regulate the range of the measured signals during logging. Figure 10.2 shows the functional scheme of the borehole device.

Connection of the borehole device with the ground remote is realized by a solid logging cable, which is also used to supply power. The borehole device works with time division into seven beats (in a regime of continuous generation of information signals). Synchronization of the information received by the ground device is realized in the moment of pause, when the information transfer from the borehole device is implemented [*Astrakhantsev and Trojanov*, 2012].

In selecting the sensors for the device for three-component geoacoustic research, we assumed that the use of a high-sensitivity accelerometer with low-frequency range, weight, and dimensions is most suitable. But the requirement for high sensitivity contradicts the requirement for low weight and a wide frequency range. The most appropriate sensors are the DN-3 type sensors [*Trojanov, Talankin, et al.*, 2014], or sensors with similar dimensions, sensitivity, and weight.

The piezoelectric accelerometer used in the device is an electro-mechanical converter, which produces the electrical output signal under the action of

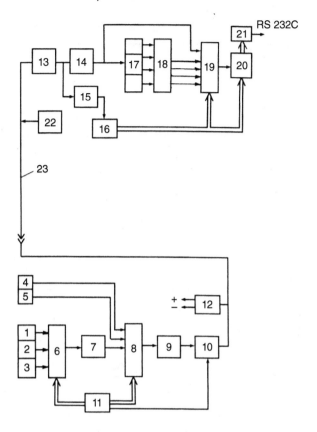

Figure 10.2 Functional scheme of the borehole device for providing three components of geoacoustic signals. Legend: 1, 2, 3: geoacoustic signal sensors; 4: temperature sensor; 5: gamma logging block; 6: first switch; 7: amplifier; 8: second switch; 9: pulse frequency modulator; 10: controlled output stage; 11: control block of the borehole device; 12: voltage regulator of the borehole device; 13: pulse former; 14: low-frequency filter; 15: pause selection block; 16: management block; 17: band filter block; 18: eliminator block; 19: third switch; 20: analog-to-digital converter; 21: shift register; 22: borehole tool power unit; 23: solid logging cable, where the output device information is registered by the PC through the PC serial port of the interface standard RS23CC.

vibration. The value of the electrical output signal level is directly proportional to the sensor oscillation acceleration in the limited frequency and dynamic range [*McKinley and Bower*, 1979; *Zeltsman*, 1968].

   The dynamic range of the accelerometer is defined as the range in which the electrical output voltage is directly proportional to the acceleration of the oscillated surface on which it is placed. The lower limit of the dynamic range is limited by noise from the measuring system [*Trojanov, Astrakhantsev et al.*, 2001; *Trojanov,*

Astrakhantsev et al., 2000]. This noise is created by the preamplifier and in some cases by the increasing temperature of the borehole device's electronics scheme, resulting from work at great depths. The upper limit of the accelerometer's dynamic range when measuring continuous oscillations is defined only by its structural strength [*Trojanov, Astrakhantsev et al.*, 2000].

The sensitivity of the piezoelectric sensor-accelerometer is defined by the ratio of its electrical output signal to the input oscillatory acceleration. For the device used, this is the ratio of the signal value in microvolts to the acceleration in mm/s^2.

10.4. Method of Measurements

The recording of signal is achieved in discrete points. The choice of the logging step depends on the problem that we have to solve, and can vary from 1 to 100 m. The survey cycle of the sensors is equal to 2 s. For each point, we arrange 20 recording cycles of the whole mass data. The total time of the device being positioned for each observation is about 30 s. The information can be seen in three windows of different colors. The first window contains the values of the GAE parameters measured at that point, the second window is the average value of the 20 measurements at each point, and the third window is used for indicating the motion of the borehole device from one measurement point to another along the borehole well. On the monitor we also record the values of the natural radioactivity and the depth of the recording point.

The geoacoustic signals are registered by three orthogonally positioned sensor-accelerometers, which measure the acceleration in mm/s^2 in different frequency bands. For the different bands, the measured parameters are distributed as follows [*Vilchinskaja and Nikolaevskij*, 1984]:

X1, Y1, Z1: signals from the horizontal and vertical sensors in the range 100–500 Hz

X2, Y2, Z2: signals from the horizontal and vertical sensors in the range 500–5000 Hz

X3, Y3, Z3: signals from the horizontal and vertical sensors in the range 500–500 Hz

X4, Y4, Z4: signals from the horizontal and vertical sensors in the range 2500–5000 Hz

Thus, in the borehole at the given depth, the values of the signals in three directions are recorded, which gives us the opportunity to compare their amplitudes in different frequency ranges. After completing the measurements for the whole observation interval, the information obtained is written to the text file, after which we perform a recalculation of the values of the measured parameters and calculate additional informative parameters, taking into account the calibration coefficients, averaging over 20 measurement cycles. The output information is given as LAS-files.

10.5. Results of Three-Component Geoacoustic Measurements Processing

The qualitative interpretation of the geoacoustic values is based on analysis of the informative measured and calculated parameters [*Trojanov and Beloglazova, 2003*]. Because the horizontal sensors X and Y are orthogonal to each other, the amplitudes of the geoacoustic signals X_n and Y_n are transformed to H_n, for each frequency range n, using formula (1):

$$H_n = \sqrt{X_n^2 + Y_n^2},\qquad(10.1)$$

where $n = 1,2,3,4$.

To reveal the casing flows, we use parameters that reflect the fluid motion in the case space: Z_n and G_n, where G_n (2) is the ratio of the vertical signal value to the horizontal value (1):

$$G_n = \frac{Z_n}{H_n}\qquad(10.2)$$

Since the three sensor-accelerometers are orthogonal to each other, the amplitude of the signal from the sensor Z in the homogeneous isotropic medium is equal to 60%–70% of the signal amplitudes from the horizontal sensors. If that relation increases due to the signal on the vertical sensor, this is an evidence of the existence of vertical fluid and gas motion processes and it is marked by anomalies Z_n and G_n.

The vertical fluid motion can have a different velocity along the borehole well, marked by spatial changing of the amplitude values of parameters Z and G. If the value of parameter G is larger than 0.8, vertical fluid flow may exist and can be estimated quantitatively by dividing the estimation into three criteria: weak, restrained, and intense. To locate the intervals of horizontal fluid and gas motion, we use the parameters H_n and M_n, where M_n is the ratio of the values from the two horizontal sensors for all frequency ranges n:

$$M_n = \frac{X_n}{Y_n}\qquad(10.3)$$

Because the sensitivity of the horizontal sensors is equal, the value of the parameter M, if the motion of fluid and gas is absent in the horizontal direction, is close to (1 ± 0.1). Deviations of the parameter M value from $1 \pm (0.4$–$0.6)$ and more are an evidence of the existence of a subhorizontal fluid motion along the layer and its intensity. The content of the revealed flow (water, oil, or gas) can be estimated on a qualitative level using the values of the group of calculated parameters N_h (4) and N_z (5). If the values of these parameters differ from the background values, there exists fluid:

$$N_h = \frac{H_2}{H_1} \qquad\qquad (10.4)$$

$$N_z = \frac{Z_2}{Z_1} \qquad\qquad (10.5)$$

So, if gas or fluid with gas (oil) exists in the medium, in the amplitude frequency spectrum a high-frequency zone of geoacoustic signals is formed (0.5–5 kHz), for which there are anomalous values of parameters H_2 and Z_2. The same anomalies can occur by a common amplitude rise for the whole spectrum in the case of registration of large signals, and these anomalies will be false. The use of parameters N_h and N_z allows us to estimate the real contribution of high-frequency signals in the GAE spectrum [*Trojanov, Astrakhantsev et al.*, 2009; *Trojanov, Astrakhantsev et al.*, 2005].

Thus, to solve the problems of revealing vertical flows and defining the intervals of horizontal fluid flow along the layer, and moreover to estimate the gaseous factor of the medium and the character of the reservoir saturation, we use separate groups of measured and calculated values of GAE parameters. Optimization of the complex of informative GAE parameters is based on the rational choice of these parameters, which to a great degree provide the solution to the given problem.

The results are represented by diagrams of the measured and calculated values of geoacoustic signal parameters.

10.6. Discussion of the Hardware and Software, Using Results for Borehole Research on the Hydrocarbon Deposits

The problems that can be solved with the borehole method, using the informative GAE parameters and further qualitative interpretation, are as follows:

The existence of amplitude anomalies in the frequency range of 100–500 Hz (parameters H_1 and Z_1) is characterized by fluid motion in the vertical and horizontal directions and also by a dynamic activity in the thickness of the sedimentary rocks.

The amplitude anomalies of high-frequency acoustical signals (parameters H_2 and Z_2) reflect the existence and motion of fluid with the gas factor, and thus the increase of the value of the vertical component signal Z compared to the horizontal is an evidence of the vertical motion of the gas-fluid mixture;

The calculated parameters $G = Z/H$, together with other parameters of acoustic signals, allow us to define the intervals of the wall fluid flows [*Trojanov and Astrakhantsev*, 2009];

If the flow motion of the oil or gas-fluid mixture along the layer is weak, when intense amplitude anomalies for the signals of the horizontal sensors are not seen, the use of the parameter $M = X/Y$ for different frequency bands allows us to fix that motion, if X/Y is not equal to unity.

If the value of the ratio $N_h = H_2/H_1$ $(N_z = Z_2/Z_1)$ is greater than 0.8, this indicates the existence of oil with a gas factor behind the casing [*Trojanov, Astrakhantsev et al.*, 2005].

So, for example, in one of the boreholes of the Astrakhan gas condensate deposit, three-component geoacoustic research was conducted to define the intervals of the casing fluid motions and possible man-made deposits, as well as for estimation of the existence of underground equipment, etc.

Analysis of the data obtained showed the following. According to the values of the horizontal parameters, anomalies of large and middle intensity were revealed at three intervals: 200–600, 1600–2000, and 2400–3600 m. Because in these intervals there had been no perforations, these anomalies are probably linked with the fluid (gas-fluid mixture) accumulation in the casing space (Figure 10.3). According to the vertical parameters, anomalies of high and middle intensity are revealed mainly in the bottom holes part of the well bore, where perforation has taken place.

According to the geological data, that interval pertains to the intervals of the Sakmar Artinskiy and Filippovsky deposits, which are the origin of the fluid flow (mineralized water + gas-fluid mixture). The increasing trend of the parameter Z from the downhole to the wellhead is an evidence of the upward motion of the gas-fluid mixture through microdefects of the underground equipment in the casing space, and this is a result of the occurring man-made hydrocarbon deposits, which are seen from the horizontal parameter anomalies.

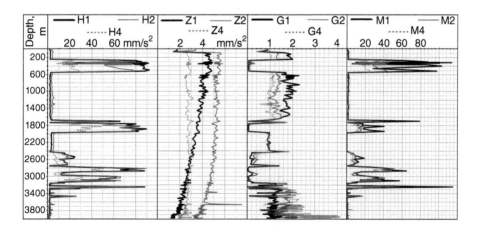

Figure 10.3 Standard complex of calculated and measured parameters from three-component geoacoustic measurements, revealing the casing flow (according to materials from OOO PKF "Nedra-S").

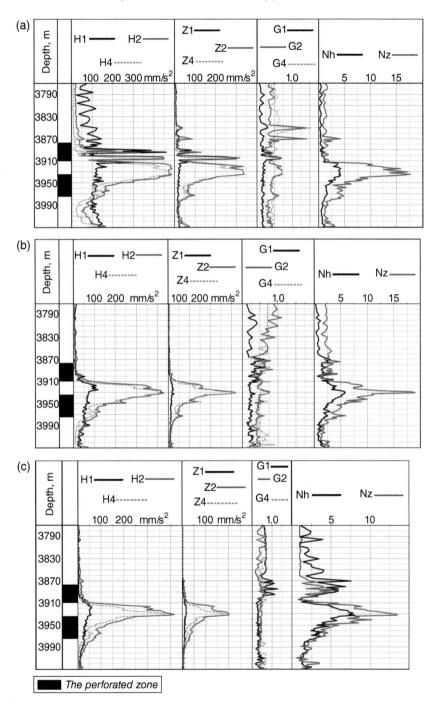

Figure 10.4 Definition of the input profile in a borehole at the Astrakhan gas condensate deposit (according to materials from OOO PKF "Nedra-S").

In the other borehole of the Astrakhan gas condensate deposit, research was carried out using geoacoustic signals together with thermal measurements (Figure 10.4) to define the inflow profile of the gas-water release intervals of the productive horizon. To solve that problem, three-component geoacoustic measurements were performed during the three-borehole working regime. Figure 10.4a presents the measurements of geoacoustic signals in a working borehole, Figure 10.4b is immediately after finishing the borehole work, and Figure 10.4c is 2 hours after finishing the borehole work. The maximum depth reached by the borehole device was 3998 m. The upper perforation intervals, pertaining to the Prikamsky and North Keltmensky horizon of the Bashkiria tier (3878–3910 m; 3935–3973 m) were investigated at the depth of 3968 m; no significant deviation from the background values of the geoacoustic signals was revealed, which indicates the weak vertical motion of mineralized water. This fact probably indicates the input of mineralized water from the interval, which was not investigated by the GAE research.

At the 3878–3910 m interval, intense anomalies of the geoacoustic signals parameters were revealed, which are an evidence of the gas input through the perforation into the casing borehole space. The second perforation interval (3935–3973 m), which pertains to the North Keltmensky horizon, shows a very weak fluid with gas factor. This fact is indicated by the effective absence of anomalies in the geoacoustic signals parameters. It must be noted that the 3943–3973 m interval practically does not work.

10.7. Conclusions

In this chapter, the effectiveness of three-component geoacoustic research is shown by dividing vertical and lateral fluid motions, which makes it possible to reveal and research the casing, interlayer, and intralayer flows of different contents and origin in the hydrocarbon deposits.

The limits of the frequency ranges for geoacoustic emission from the motion of water, fluid with gas factor, and gas were determined. The possibility of defining the current reservoir saturation was considered by using this method.

The possibilities of three-component (oil, gas, and gas condensate deposits) measurements using the geoacoustic signals method expand the boundaries of the given method's use, contribute to the development of the complex geophysical exploration, and increase its informativity.

Thus, the use of three-component measurements of geoacoustic signals from hydrocarbon deposits makes it possible to solve a wide spectrum of problems through the control of oil and gas borehole exploration:

1. Revealing the fluid motion intervals for the layers/reservoirs that have not been opened by perforation
2. Defining the character of saturation of the layers/reservoirs
3. Defining the casing flows in cases where the interpretation of standard thermal data is not unique

4. Research of the fluid input profile and the fluid content in the intervals of layer/reservoir perforation through a static and dynamic borehole working regime
5. Revealing and detailing the locations of borehole leaks (together with the thermal data)
6. Revealing disintegrated areas in outer borehole spaces

Acknowledgments

The authors would like to express their appreciation to A. K. Trojanov. The obtained results include his methodological ideas for the research of oil-gas boreholes.

The work was fulfilled with financial support from the Government of Sverdlovsk region and the Russian Fund for Fundamental Research: project RFBR- Ural No. 13-05-96019 and also project RAS 14-5-NP-214.

References

Astrakhantsev, G., and A. Trojanov (2012), *Device for providing geoacoustic logging*. Patent RF, 2445653, Bulletin, 8 (in Russian).

Dobrinin, V. (1965), *Physical properties of oil gas reservoirs in deep boreholes*. Moscow, Nedra (in Russian).

Dobrinin, V, Gorodnov, A. et al. (2001), Experience of using the processing and interpretation technology of acoustic logging for research of oil and gas boreholes. *Geophysics*, 4, 58–65 (in Russian).

Ivanov, B. (2010), *World of physical hydrodynamics*, 2nd ed. Editorial URSS, Moscow (in Russian).

McKinley, R., and F. Bower (1979), Specialized applications of noise logging. *J. Pet. Tech.*, 1387–1395.

Tiab, D., and E. Donaldson (2009), *Petrophysics: Theory and practice of measuring reservoir rock and fluid transport properties*, 2nd ed. Premium Engineering, Moscow.

Trojanov, A. (2000), Research of present day dynamic state of geological medium, according to borehole observations of geoacoustic noise. *Ural Geophysical Bulletin*, Ekaterinburg, UB RAS, 1, 88–94 (in Russian).

Trojanov, A., and Y. Astrakhantsev (2009), *Method of revealing casing fluid flows in the boreholes*. Patent RF 237 3392, Bulletin 32 (in Russian).

Trojanov, A., Y. Astrakhantsev, et al. (2000), Three-component geoacoustic logging-methodical and device-programmed software. *Karotazhnik*, 68, 17–33 (in Russian).

Trojanov, A., Y. Astrakhantsev, et al. (2001), Hardware-software system for three-component geoacoustic logging and examples of problems solution by control of the oil deposits exploitation. *3-D Congress of Oil Owners of Russia. Scientific Symposium, New technologies in geophysics*, Abstracts, Ufa, edited NIIBZHD, pp. 29–30 (in Russian).

Trojanov, A., Y. Astrakhantsev, et al. (2005), *Method of defining the character of layers-reservoirs saturation*. Patent RF 2265868, Bulletin 34 (in Russian).

Trojanov, A., Y. Astrakhantsev,et al. (2009), *Method of revealing gas saturated layers in boreholes*. Patent RF 2344285, Bulletin 2 (in Russian).

Trojanov, A. and Beloglazova, N. (2003), Optimization of the system of geoacoustic signals informative parameters by solving the problems in oil and gas boreholes. *Proc. XIII Session of the Russian Acoustic Society*, Moscow, pp. 57–60 (in Russian).

Trojanov, A., G. Igolkina, et al. (2012), Possibilities of three-component geoacoustic logging by control exploitation of the oil deposits. *Geophysics*, 1, 36–41 (in Russian).

Trojanov, A., A. Talankin, et al. (2014), *Use of three-component geoacoustic logging for exploitation of gas and gas-condensate deposits*. Tutorial, Karaganda (in Russian).

Viktorov, I. (1981), *Sound surface waves in solid medium*. Nauka, Moscow (in Russian).

Vilchinskaja, N., and V. Nikolaevskij (1984), Acoustic emission and spectra of seismic signals. *Izvestija AN USSR, Fizika Zemli*, 5, 91–100 (in Russian).

Vorobjev, A., and M. Tonkonogov, et al. (1972), *Theoretical questions of rock physics*. Nedra, Moscow (in Russian).

Zeltsman, P. (1968), *Development of devices for geophysical borehole researches*. Nedra, Moscow (in Russian).

11. APPLICATION OF BOREHOLE MAGNETOMETRY TO STUDY OIL AND GAS DEPOSITS IN WESTERN SIBERIA

Galina V. Igolkina

Abstract

Lithological decomposition of the borehole section is based on differences of the values and character of the curves of the rock magnetic susceptibility, magnetic field, and magnetization. Magnetic features of the rock are reliable criteria for lithological decomposition of thin-layered, heterogenic sections of deep oil-gas boreholes and also for providing control of the geological documentation of the borehole sections. A method is developed for identification and correlation of magnetic rock inside the borehole space. The magnetic parameters that are used for providing correlation are magnetic susceptibility; magnetic field; the whole, inducted, and residual magnetization; the value and the sign of the factors Qz and Qh and others. The interpretation of the results of borehole magnetometry allows establishment of the existence of two types of magnetic mineralization: pyrothine, which is linked with the genesis zones of sulfur (pyrite) mineralization, and magnetite (titan magnetite), which has been confirmed by the results of pale magnetic research of the core. The examples of the use of the results of borehole magnetometry for constructing a volume model for the near borehole area can be the use of magnetic rock correlation along the oil gas boreholes of the Siberian platform and western Siberia. The use of that method increases the significance of structural constructions of geological data and provides the possibility of deep structural forecast. The deep

Institute of Geophysics Ural Branch of Russian Academy of Sciences (UB RAS), Ekaterinburg, Russia

Oil and Gas Exploration: Methods and Application, Monograph Number 72,
First Edition. Edited by Said Gaci and Olga Hachay.
© 2017 American Geophysical Union. Published 2017 by John Wiley & Sons, Inc.

structural forecast is mainly linked with the quality of rock correlation in the inner borehole space. Therefore, the use of borehole magnetometry is of a decisive significance. The research of oil gas complexes and intrusions that occur in sedimentary basins is needed for revealing the regularities of oil-gas deposit locations and their right forecast.

11.1. Introduction

A thorough study of oil and gas complexes and intrusions that occur in sediment basins is needed for formulating the patterns of oil and gas deposit locations and their accurate forecasting. Borehole magnetometry allows researchers to obtain a set of magnetic characteristics for defining the layers-collectors beddings. The link between these geological objects and geophysical methods of borehole studying sections is petrophysics. The advantage of borehole magnetometry lies in the discovery of the rock in situ, which is of large significance by drilling oil and gas boreholes low core recovery.

Borehole magnetometry is successfully used to research deep wells in the oil and gas regions in Russia [*Glukhikh et al.*, 1995; *Igolkina*, 1995; 2002]. Therefore, research in oil and gas boreholes demanded a new, more sensitive, precise, high-temperature logging tool, using a new processing and interpretation method [*Astrakhantsev, Beloglazova*, 2012; *Astrakhantsev, Beloglazova et al.*, 2008; *Igolkina et al.* (b), 1999].

Borehole magnetometry lends the ability to procure a complex of magnetic characteristics. Thus, it gives us the opportunity to provide with a sufficient degree of validity a comparison of magnetic rock located around super deep boreholes [*Astrakhantsev, Beloglazova et al.*, 2008; *Glukhikh et al.*, 1995; *Igolkina*, 2002]. The successful solution of the problem for each borehole depends on concrete geological conditions and physical rock features, defined from core measurements in a natural location.

We started with the problem of extraction, identification, and correlation of trap intrusions. For its solution, we used the deep magnetometrical structural forecast data from the northwestern part of the Siberian platform [*Igolkina and Svyazhina*, 1995; *Igolkina*, 2002].

The results of research of super deep, deep, and oil and gas boreholes show that using diagrams of magnetic susceptibility and magnetic field together with the data of other methods of GIS and petro physical information, we can define and extract the intervals of magnetic rock, and estimate the magnetic features, section lithology, and rock heterogeneity.

The connecting link between the geological objects and geophysical borehole sections is petro physics. Some questions of geological interpretation of the data can be answered by the visual analysis of the diagrams of magnetic susceptibility and of the magnetic field components. We can extract different types of intrusion and effusion rocks on the background of weak magnetic tuffs, aleurolite, and practically nonmagnetic dolomites and limestone using borehole magnetometry in the oil and gas boreholes [*Astrakhantsev et al.*, 2008; *Igolkina et al.*, 1999b; *Mezenina et al.*, 2004].

Defining rock magnetization in natural occurrences; estimating its value, sign and change with depth; and also researching the correlation between the magnetic field and magnetic susceptibility for different types of rock has allowed estimation of mineralization types using oil and gas boreholes [*Igolkina*, 2002]. The peculiarities of the rock magnetization for each borehole have been researched and the comparison has been provided. These obtained results give the opportunity to use them for geological interpretation of the inner and outer magnetic field. The advantage of borehole magnetometry and being able to research rock in its natural occurrence is of great significance because only minimal data is obtained from the core in drilling oil and gas boreholes.

11.2. Method of Borehole Magnetometry

Devices to measure constant magnetic field in deep boreholes were developed at the Institute of Geophysics UB RAS. These devices allow researchers to make important measurements of the magnetic susceptibility of rocks (χ), the vertical component (Z_a) and the modulus of the horizontal vector (H_a) of the geomagnetic field, the magnetic azimuth ($A_з$), and the zenith angle (ϕ) of the borehole. The device MI-6404 is thermal pressure resistant up to 250°C, 220 МПа. All measurements are fulfilled during two up and down operations using a three-wire logging cable [*Astrakhantsev and Beloglazova*, 2012]. The methods of processing and interpreting borehole magnetometry results from deep boreholes is described in the book by G. *Igolkina* [2002]. That interpretation of magnetometry discovery is needed for extracting the geological information from the data of magnetic rock features in situ. Its main problem consists in defining the peculiarities and changes in the rocks' magnetization in deep oil and gas boreholes. It is based on the connection between magnetic anomalies and following geological factors: lithological rock type, degree of their change, structural and textural peculiarities, and type and concentration of magnetic minerals. This connection is understood through constructing an oil and gas deposit model.

11.3. Results and Discussion of the Research

The division of the well section is based on differences in the magnitude and character of the magnetic susceptibility and the rock's magnetic field in the oil and gas wells. Magnetic rocks from deep boreholes in the sedimentary regions possess different magnetic properties and a less clear magnetic differentiation with the surrounding rocks than is observed for more strongly magnetic rocks. For weak magnetic objects, the ratio of magnetic susceptibility χ between the magnetic body and imbedded rocks does not exceed 1–2 orders of the value. The more homogeneous according to their magnetic properties are sedimentary rocks such as clay, aleurolite, mudstone, sandstone, limestone and dolomite. Their magnetic susceptibility varies within comparatively large limits, 0–400 × 10^{-5} units of SI, but does not usually change very sharply from point to point, which is a favorable factor when using a magnetic correlation method on the oil and gas deposits.

The magnetic properties of the sedimentary rock complex are stipulated by accessory ferrimagnetic minerals: magnetite and its variants maghemite, hematite, and iron hydroxides, pyrothine and so on. The rock-forming minerals of the sedimentary rocks, quartz, calcite, feldspath, gypsum, anhydrite, and halite, are diamagnetic or weak paramagnetic rocks and do not contribute much to the value of the rocks' magnetic susceptibility. Among the paramagnetic minerals, the main role is played by siderite, chlorite, pyrite and sometimes clay minerals, whose role is stipulated by impurities, relicts, and new formations of iron oxides of minerals [*Khramov*, 1976].

Practical methodical works for the selection of the complex geophysical research, to solve the problem of the lithological division of the sedimentary cover and Paleozoic basement of western Siberia, were carried out in boreholes NX1 and NX2 of the Kechimow deposit and in borehole NX3 of the Tevlin-Russkinsk deposit [*Igolkina et al.*, 1999b].

Analysis of the results of borehole magnetometry shows that the sedimentary rocks in these boreholes are weakly magnetic and the curves of the magnetic susceptibility χ of the magnetic field are highly differentiated, which corresponds to the known interstratifications of mudstones, sandstones, and aleurolite with different magnetic properties. The differences in the magnetic susceptibility values are linked with the heterogeneous distribution and amount of magnetic minerals in the rock. The sandstones have low values of magnetic susceptibility and magnetic field; the value of magnetic susceptibility does not exceed 80 × 10^{-5} units SI. Aleurolite and argillites have values of magnetic susceptibility χ up to 150 × 10^{-5} units SI. Narrow layers of coal, which are not practically magnetic, can occur there.

An interpretation of the borehole magnetometry results allows us to identify the existence of two types of magnetic mineralization: pyrothine, which is linked with evolution zones of sulfide (pyrite) mineralization, and magnetite (titanium-magnetite), linked with basalts, which was proven by the results of core samples research [*Igolkina*, 1985, 2002; *Igolkina et al.*, 1999b]. Pyrrhotite mineralization, according to borehole magnetometry data in borehole NX1 of the Kechimow deposit, is located at the intervals 2662–2667 m, 2689–2690 m (according to the geological description of the silt, pyritization of the rocks was not registered), and also at depths of 2849–2850 m and 2858–2868 m. According to the geological description of the silt, the presence of pyrite was registered at the intervals 2844–2847 m and 2853–2867 m, and according to description of the core samples in the intervals 2846–2847 m and 2857–2862 m. Research on the magnetic susceptibility of the core samples from borehole NX1 showed that magnetic basalts, for which the value of magnetic susceptibility is up to 4500×10^{-5} units SI, are registered at depths of 3463–3465 m, while weathered basalts are located from the depth of 3462 m. The increased values of χ up to 1000×10^{-5} units SI are registered also for porphyrite at the depth of 3540.5 m and basalts at the depths 3548–3549 m according to borehole NX1. Massive and weathered basalts were registered in borehole N153 at the interval 3461–3468 m; the value of their magnetic suscepti-bility is up to 600×10^{-5} units SI, and a magnetic field anomaly of Z_a up to 400 nT was registered. Lower down the narrow nonmagnetic limestone layer, magnetic basalts were registered at depths of 3477–3482 m, for which the value of $\chi = 2500 \times 10^{-5}$ units SI, and the value of Z_a is from 500 nT to 1000 nT. According to the core samples, the effusive rocks (dark-green basalts with red undertone) were registered in the intervals 3456–3463 m, 3483–3486 m, 3540–3550 m and the value of magnetic susceptibility was 800×10^{-5} units SI.

According to borehole NX2, the research of metamorphic argillite's magnetic susceptibility at the depths of 3375–3380 m and 3382–3389 m showed that its value is up to 600×10^{-5} units SI. Higher values of χ up to 1400×10^{-5} units SI were registered for basalts at the depth of 3486 m. The effusive rocks registered in borehole N155 are magnetic and the value of magnetic susceptibility is up to 7000×10^{-5} units SI.

Metamorphic and pyroclastic rocks (tuffs and tuff sandstones) registered in borehole NX3 of the Tevlin-Russkinsk deposit at depths of 3444–3453 m, 3521–3524 m, 3528–3529 m, 3552– 3560 m, 3593–3603 m, 3635–3640 m, 3762.5–3770 m, 3825–3832 m, and 3923–3928 m are characterized by values of magnetic susceptibility from 0 to 80×10^{-5} units SI. The magnetic susceptibility of red hematite granites, cracked and massive at depths of 3963–3967 m, varies from 210 to 500×10^{-5} units SI.

According to these measurements of magnetic susceptibility, we constructed histograms of the rocks distribution (Figure 11.1).

Moreover, we constructed diagrams comparing the χ of the core samples and the results of borehole magnetometry, which showed a good comparability and the benefit of borehole magnetometry, because only the latter yielded data from the whole section in situ as opposed to the particular intervals for which we had core samples. On the other hand, if we have sufficient information on magnetic susceptibility for the research of core samples, we can compare the data from these intervals with the integral rock models. The behavior of the curves χ can be explained by different factors, such as the hydrodynamic conditions, nonconstant and heterogeneous distribution of the magnetic minerals, the existence of inclu-sions, and structural-textural rock heterogeneities. Examples comparing magnetic susceptibility χ from the borehole magnetometry data and from core samples data are presented in Figure 11.2.

In addition, we provided paleomagnetic research of core samples registered for the Mesozoic sedimentary mantle and the weathering crust of the Paleozoic

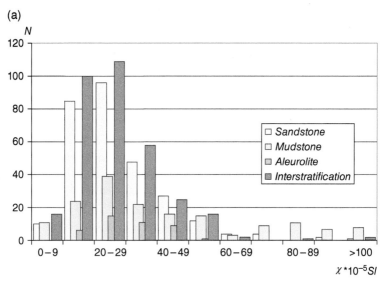

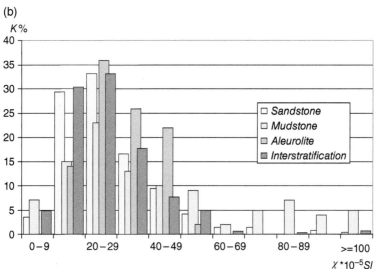

Figure 11.1 Distribution of rocks, according to their magnetic susceptibility χ, for borehole NX2. a: number of specimens (N); b: percentage content of the rocks ($K\%$).

basement to verify the results of the borehole magnetometry. As objects of magnetic and paleomagnetic research, we used the rocks of layer AB-1 of the Alim suite, low chalk, and layers SE-9 and SE-10 of the Tyumen suite, low and middle Jurassic, from borehole NX2; Mesozoic sediments without division into

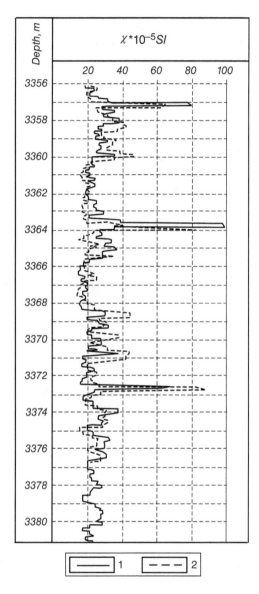

Figure 11.2 Result of comparing the distribution of magnetic susceptibility χ, according to borehole magnetometry data (1) and core samples (2) in borehole NX1.

suites from boreholes NX1 and NX3; and sediments of weathering crust and effusive rocks of the Paleozoic basement from boreholes NX1, NX2, and NX3 [*Svyazhina et al.*, 1996; *Mezenina et al.*, 2004]. Mesozoic sediments have weak magnetic properties: the value of magnetic susceptibility χ varies from 3 to 63 × 10^{-5} units SI, the natural residual magnetization J_n from 0.09 to 39.4 mA/m. The vectors J_n have steep inclinations I_0 that are directions near to the axis of the core sample. For the Mesozoic sandstones, the self-reversal J_{rs} is registered at temperature intervals of 420–600°C, which proves the existence in the rock of magnetic iron sulfides, including high-temperature phase. By heating, first the destruction of the component of magnetization occurs, which is stipulated by pyrrhotite with a temperature of 320–380°C (Curie), and in the interval 400–600°C, by the second component J_{rs} of the opposite direction, linked with super-pyrrhotite.

The magnetic properties of the Paleozoic rocks investigated, namely specimens of aleurolite, argillite, and effusions of the basement, vary in wide limits: χ from 23 to 2975 × 10^{-5} units of SI, and J_n from 8.9 to 4253 mA/m. The vectors J_n have a large spread in values of inclinations, with the dominant angles less than 45 degrees. The natural residual magnetization of the Paleozoic rock, by contrast with the Mesozoic sediments, has two components [*Svyazhina et al.*, 1996; *Mezenina et al.*, 2004].

The method developed at the Institute of Geophysics UB RAS for defining the complex J, inductive J_i and residual J_n magnetizations of heterogeneous media using the data of the inner magnetic field and magnetic susceptibility allow us to research the magnetic properties of the rock in situ. Comparison of the magnetization calculations with the results of defining the residual magnetization of the core samples shows good agreement between the results [*Igolkina*, 2007, 2002; *Igolkina and Svyazhina*, 1995].

The use of borehole magnetometry for the correlation of geological sections along the deep and super-deep boreholes demands a special approach for each borehole and working region. With this aim, a complex of indicators for magnetic correlation was developed, which allows us a sufficient degree of validity to provide rock comparisons along deep boreholes. They are the value of magnetic susceptibility, the value and the sign of the magnetic field, the value and the direction of the natural residual magnetization, which are defined from the borehole measurements, and also other magnetic characteristics [*Igolkina*, 2002].

The correlation of effusive rocks was fulfilled using sections NX1 and NX2 of the boreholes of the Kechimow deposit of western Siberia. Interpretation of the borehole magnetometry data shows that the cracked, weathered basalts are characterized by values of magnetic susceptibility up to 600 × 10^{-5} units SI and by the anomalous magnetic field Z_a up to 400 nT, according to borehole N153 at depths of 3462–3468 m, while in borehole N155 lower values are registered at depths of 3473–3487 m (Figure 11.3) [*Igolkina et al.*, 1999b].

The magnetic basalts registered in the boreholes have values of magnetic susceptibility from 2500 to 3800 × 10^{-5} units SI and values of the anomalous vertical

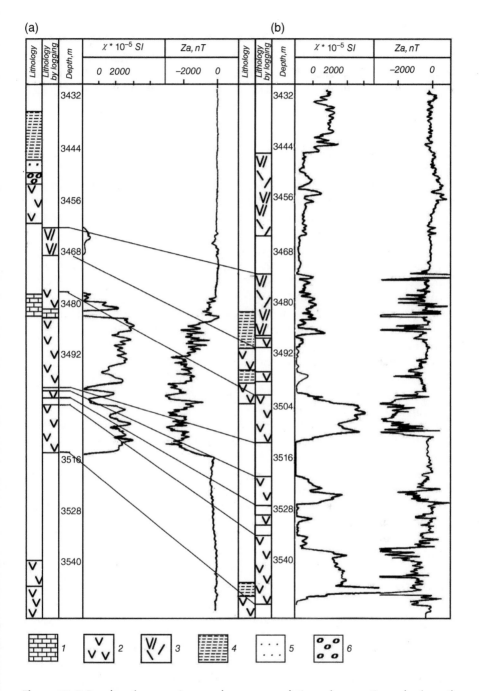

Figure 11.3 Results of comparison and space correlation of magnetic rocks from the Kechimow deposit, according to data from boreholes NX1 (a) and NX2 (b), at depths of 3330–3550 m. Legend. 1: limestones; 2: basalts; 3: cracked basalts; 4: argillites; 5: sandstones; 6: conglomerates.

component of the magnetic field up to 3000 nT. The anomaly of the module of the horizontal component H is about 6000 nT. The curves of magnetic susceptibility χ and Z_a are highly differentiated. The character of the curves Z_a and H corresponds to the fact that the basalt bodies are transversal to the imbedded rocks. The definition of those layer attitude elements according to the magnetic field steps proves that the southwest rocks dip. Basalts with a thickness of 38 m (depths of 3477–3514 m) with thin stringers of sedimentary rocks are registered in borehole N153, while in borehole N155 they are registered among the sedimentary rocks at depths of 3501–3512 m, 3520–3526 m and 3537–3549 m with the same magnetic properties (Figure 11.3) [*Igolkina et al.*, 1999b].

The magnetic rocks similar to metamorphic rocks (argillites), which are registered in borehole NX2 according to magnetic borehole data at depths of 3352–3464 m were not registered in borehole NX1. Thus, the correlation scheme shown in Figure 11.3 allows us to improve the reliability of the geological-geophysical section of the oil deposit.

Besides that, the results of borehole magnetometry show that anomalies can be registered in the boreholes when measuring the magnetic field and χ; these are linked with metal in the space around the borehole and also with metallic particles in the walls of the borehole and cavities. The existence of the drilling fluid with higher magnetic properties, adsorption on cavities in the borehole walls and on absorbing zones, as well as the existence of magnetic particles can be used for allocation in the borehole sections with cracks and disintegration zones [*Igolkina*, 2013; *Igolkina et al.*, 1999a].

Research on the magnetic characteristics of the iron concentrate IC-1 and drilling fluids, weighted by IC-1, was carried out to determine the influence of the weighted IC-1 on the results of the magnetometric borehole research data [*Igolkina et al.*, 1999a, b]. Research was conducted on the magnetic susceptibility and the magnetic field of the drilling fluid, weighted by IC-1.

Systematic investigation to delineate the boundaries of the oil reservoir and crack zones using sections of oil boreholes was performed by the device MI-3803 using magnetic fluid as proposed by the author [*Astrakhantsev et al.*, 2008; *Igolkina*, 2013, 2002]. The content of that method is based on research on the magnetic properties of the rocks before and after processing the borehole with magnetic fluid IC-1 (two-fluid method).

Interpretation of the borehole magnetometry data, obtained from borehole NX3 of the Tevlin-Russkinsk deposit, filled by the weighted magnetic IC-1 drilling fluid with specific weight of 1400 kg/m³, allowed us to distinguish permeable cracked zones in the section. The iron concentrate IC-1 is a mixture of iron oxides with small additions of titanium and vanadium. It is a dark grey or black thin dispersed powder with particle dimensions of 0.001–0.1 mm and density of 4600–5000 kg/m³ [*Igolkina et al.*, 1999a].

The crack zones, which are seen on the background of the magnetic drilling fluid with magnetic susceptibility about 20,000 × 10^{-5} units SI, can be registered

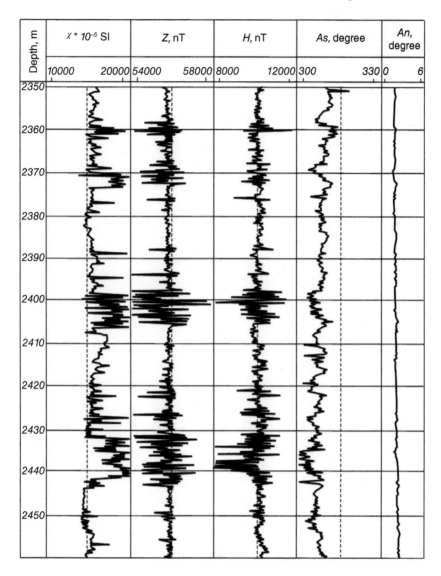

Figure 11.4 Research of crack zones in borehole NX3 of the Tevlin-Russkinsk deposit.

by anomalies of magnetic susceptibility χ and by magnetic fields components: vertical component Z and the module of horizontal component H, the magnetic azimuth of the borehole As, and the zenith angle of the borehole An. The anomaly zones are linked with infiltration into the pores and cracks of thin powdered magnetite, which is the basis of the iron concentrate IC-1 (Figure 11.4).

11.4. Conclusions

Analysis and interpretation of the results of borehole magnetometry and magnetic and paleomagnetic research on core samples from boreholes NX1 and NX2 of the Kechimow deposit and borehole NX3 of the Tevlin-Russkinsk deposit allowed us to solve geological problems such as selection in the borehole sections of rocks with higher magnetic properties; clarification of their thicknesses; estimation of the magnetic characteristics of rocks, along with their polarity and sizes; determination of the spatial positions of these rocks; and determination of the rocks' magnetic susceptibility in situ. Correlation of magnetic rocks is executed in the internal downhole space and confirmed by the high degree of correlation between data obtained using different GIS methods. It is possible to estimate fractured and cracked zones in the sections of the oil and gas boreholes. A deep structural prognosis is, to a great extent, related to the quality of rocks' correlation in the interwell space and, in this respect, the application of borehole magnitometry is of decisive value.

The magnetic properties of rocks are a reliable criterion for lithological division of the thin- layered, heterogeneous oil and gas sections.

Acknowledgments

The work was fulfilled with partial financial support from the Russian Academy of Sciences, Department of Earth Sciences, and fundamental research.

References

Astrakhantsev, Y., et al. (2008), Magnetometrical survey of oil and gas wells. *Exposition Oil and Gas*, 4, H(69), 12–14 (in Russian).

Astrakhantsev, Y., and N. Beloglazova (2012), *Integrated magnetometrical equipment for research and ultra-deep exploration wells.* Ural Branch of Russian Academy of Sciences, Ekaterinburg (in Russian).

Glukhikh, I., et al. (1995), Magnetometry of ultra-deep and deep wells. *Geophysics*, 4, 37–41 (in Russian).

Igolkina, G. (1985), *The use of downhole magnetic survey to study trap intrusions: Algorithms, methods and results of geophysical data interpretation.* Naukova Dumka, Kyiv (in Russian).

Igolkina, G. (2002), *Borehole magnetometry in investigation of deep and super-deep boreholes.* Ural RAN, Ekaterinburg (in Russian).

Igolkina, G. (2007), Study of magnetization of rocks in situ by measurements in ultra-deep and deep wells. *Vestnik MSTU, Proceedings of Murmansk State Technical University*, 10(2), 244–250 (in Russian).

Igolkina, G. (2013), Solution of technological problems using magnetometry data in ultra-deep oil and gas wells. *NTV, Karotazhnik*, 230, 25–40 (in Russian).

Igolkina, G., and I. Svyazhina (1995), Correlation of borehole magnitometry results with paleomagnetic investigation on cores at Muruntau SD-10. *Book of Abstracts, XXI General Assembly of IUGG, Scientific Program GA 5.19, "Magnetic Petrology and Magnetic Signature of Ore Deposits and Ore Environments,"* 2–14 July 1995, Boulder, CO, USA.

Igolkina, G., et al. (1999a), Study of the magnetic properties of iron-ore concentrate weighting IC-1. In *Electrical and electromagnetic methods of research in oil and gas wells.* SB RAS, SIC UIGGM, Novosibirsk, pp. 298–304 (in Russian).

Igolkina, G. et al. (1999b), Prospects and opportunities in the study of borehole magnito-metrical data for Paleozoic sedimentary sections of the foundation of western Siberia, measured in deep wells. In *Electrical and electromagnetic methods of research in oil and gas wells.*, RAS, SIC UIGGM, Novosibirsk, pp. 291–298 (in Russian).

Khramov, A. (1976), Magnetic properties of rocks. In *Physical properties of rocks and minerals.* Nedra, Moscow, pp. 185–186 (in Russian).

Mezenina, Z., et al. (2004), Study of magnetic and paleomagnetic core specimens' prop-erties of oil boreholes in western Siberia. *Book of Abstracts: 5th International Conference on Problems of the Geocosmos, St. Petersburg,* 179–180 (in Russian).

Svyazhina, I., et al. (1996), Diagnosis of magnetic iron carbides in the carbon-micaceous schists of Muruntau SDB. *DAN,* 347(6), 792–794 (in Russian).

12. A NOVEL MODEL TO ESTIMATE S-WAVE VELOCITY INTEGRATING HÖLDERIAN REGULARITY, EMPIRICAL MODE DECOMPOSITION, AND MULTILAYER PERCEPTRON NEURAL NETWORKS

Said Gaci

Abstract

The prediction of S-wave velocity (Vs) is an important issue in the characterization of oil and gas reservoirs. Due to the complexity of this task, the traditional estimating methods often fail to capture the nonstationary and nonlinear properties of this parameter and to accurately quantify it. In this study, a prediction model integrating Hölderian regularity, empirical mode decomposition (EMD), and a multiple layer perceptron artificial neural network (MLP ANN) is proposed to predict Vs from P-wave velocity (Vp).

Here, the Vp and Vs *logs* are assumed to be realizations of multifractional Brownian motion (mBm) processes. It is shown that for a given borehole, the regularity logs, or the Hölder exponent profiles, computed from both velocity logs exhibit a strong correlation. The regularity is a robust attribute of lithology; thus, it can be considered as a supplementary input for neural network models.

First, the suggested model uses EMD to decompose Vp log data into a set of intrinsic mode functions (IMFs) and one residue. Next, the obtained IMFs and residue are used to construct a high-frequency (HF) component, a low-frequency (LF) component, and a trend component using the Hölderian regularity-based fine-to-coarse reconstruction algorithm. Then, the MLP ANN algorithm is used to predict Vs log from Vp log considering

Formerly, Sonatrach, Division Exploration, Boumerdès, Algeria
Now, Sonatrach IAP, Boumerdès, Algeria.

Oil and Gas Exploration: Methods and Application, Monograph Number 72,
First Edition. Edited by Said Gaci and Olga Hachay.

four combinations in inputs: (a) simply *Vp log*, (b) *Vp* and its regularity *log Hp*; (c) *Vp log* and (HF, LF and trend) components, and ultimately, (d) *Vp, Hp,* and (HF + LF + trend) components.

The MLP ANN model uses the main KTB borehole (KTB-HB) *log* data for training and testing to predict *Vs* velocity *log* from *Vp logs* recorded at the pilot KTB borehole (KTB-VB). The results show that the predicted *Vs* values using MLP ANN with the last input combination are more accurate than those obtained with the other combinations.

Therefore, the MPL ANN integrating as input *P*-wave velocity *log*, its Hölderian regularity *log*, and the three components obtained by the fine-to-coarse reconstruction algorithm is a powerful tool for predicting *S*-wave velocity.

12.1. Introduction

The prediction of *S*-wave velocity is a major goal in petroleum exploration. The investigation of the relationship between *P*- and *S*-wave velocities (*Vp* and *Vs*) allows a better understanding of petrophysics, inversion, and formation evaluation.

In literature, many equations based on *Vp* rather than on porosity have been suggested to estimate *Vs*. They are mostly empirically developed for wet sediment. *Pickett* [1963] showed that $Vs = Vp/1.9$ for limestone and $Vs = Vp/1.8$ for dolomite. Later, based on the famous "mud rock line," *Castagna et al.* [1985] gave $Vs = 0.862Vp - 1.172Vp$. Then, *Castagna et al.* [1993] made some changes to these equations: $Vs = -0.055Vp^2 + 1.017Vp - 1.031$ for limestone and $Vs = 0.583Vp - 0.078$ for dolomite, with the velocity in km/s. In addition, they suggested meanings for elastic rock: $Vs = 0.804Vp - 0.856$, using a large experimental sandstone dataset with large ranges of porosity and clay content variation, and *Han* [1986] obtained $Vs = 0.794Vp - 0.787$. These measurements were carried out on wet rock at ultrasonic frequency.

Later, *Mavko et al.* [1998] found $Vs = 0.79Vp - 0.79$ by adding to these measurements a number of data points from high-porosity unconsolidated sands. Analysis of the dataset used by *Han* [1986] considering the porosity values yields $Vs = 0.853Vp - 1.137$ for porosity below 0.15 and $Vs = 0.756Vp - 0.662$ for porosity greater than 0.15. By using well log data, *Williams* [1990] established $Vs = 0.846Vp - 1.088$ for water-bearing sands and $Vs = 0.784Vp - 0.893$ for shales.

To eliminate the inadequacy of the linear models and account for the nonlinear patterns encountered in real problems, several nonlinear models have been suggested, among which multiple layer perceptron artificial neural networks (ANNs) have shown high nonlinear modeling potential. The MLP ANNs have been widely used for the purpose of predicting lithofacies from well logs data [*Frayssinet et al.*, 2000; *Goutorbe et al.*, 2006]. Even though MLP ANNs are a powerful technique within the stationary domain, their application for predicting well logs, which are nonstationary, leads to inadequate results. In this chapter, in order to improve the prediction efficiency, additional inputs are considered in ANNs inputs: the local fractal (or regularity) property (measured by the Hölder exponent) and the nonlinearity and nonstationary of well logs analyzed using empirical mode decomposition (EMD).

Fractal geometry offers an appropriate tool for investigating the spatial heterogeneities of different geological patterns [*Mandelbrot*, 1975, 1977]. It has been successfully used in various fields of geosciences, particularly in oil and gas engineering. *Hewett* [1986] was the first to introduce statistical fractals to model well logs. Then, *Hardy and Beyer* [1994] showed that the latter can be described by fractional Brownian motion (fBm). fBm is a nonstationary stochastic fractal model with stationary increments. It is indexed by a Hurst (self-similarity) exponent $(0 < H < 1)$. Since this process exhibits the same local regularity (H) everywhere, it does not allow analysis of real signals, which present time/space-dependent regularity. This drawback is solved by allowing H to evolve in time (and/or space); the obtained process is known as multifractional Brownian motion (mBm). mBm is gaining increasing attention in various research fields: geophysics [*Wanliss*, 2005; *Wanliss and Cersosimo*, 2006; *Cersosimo and Wanliss*, 2007; *Gaci et al.*, 2010, 2011; *Gaci and Zaourar*, 2010, 2011a, 2011b, 2014a; *Gaci*, 2014a], image-processing [*Bicego and Trudda*, 2010], traffic phenomena [*Li et al.*, 2007], etc.

In previous works [*Gaci et al.*, 2010; *Gaci and Zaourar*, 2010, 2011a; *Gaci*, 2014a], sonic well logs were considered realizations of mBm processes. Then, the depth-dependent regularity profiles derived from the logs allowed us to perform a lithological segmentation and to investigate the heterogeneities of the geological layers crossed by the well. Here, we show that Hölderian regularity is a reliable property of lithology that can be considered in the input parameters for neural network prediction.

EMD is an adaptive time-frequency analysis method suggested by *Huang et al.* [1998, 1999]. It is a data-driven method that does not require any a priori assumptions on the decomposition basis and is appropriate for the analysis of nonlinear and nonstationary data. This technique has received increasing interest in numerous domains of application [*Gloersen and Huang*, 2003; *Chiew et al.*, 2004; *Huang and Wu*, 2008; *Bekara and Van der Baan*, 2009; *Han and Van der Baan*, 2011; *De Michelis et al.*, 2012; *Gaci*, 2014b,c; *Gaci and Zaourar*, 2014b]. The EMD algorithm consists of decomposing the signal into a sum of components, namely intrinsic mode functions (IMFs), displaying different local characteristics of the well log data. In fact, by considering only a well log in the ANNs inputs, the networks make time to accurately capture the nonlinear and nonstationary properties of the data and to understand their interference between the different scales. In order to enhance the ANNs' prediction efficiency in terms of time and accuracy, the IMFs representing the local characteristics of well log data at each of the scales involved in the EMD method are used.

The main contribution of this paper is to introduce additional logs, derived from *Vp log*, in the MLP ANNs algorithm to predict *Vs log*. These logs are the Hölderian regularity log and three components: a high-frequency (HF) component, a low-frequency (LF) component and a trend component calculated with the Hölderian regularity-based fine-to-coarse reconstruction algorithm.

The rest of this study is organized as follows. Section 12.2 describes the theoretical background: Hölderian regularity, EMD, Hölderian regularity-based fine-to-coarse reconstruction algorithm, and MLP ANN. Section 12.3 reports the empirical results; finally, section 12.4 presents some conclusions.

12.2. Theory

12.2.1. Local Hölderian Regularity

For a stochastic process X whose trajectories are continuous but nowhere differentiable, the Hölder exponent is defined by:

$$\alpha_X(z_0) = \sup\left\{\alpha, \limsup_{h \to 0} \frac{\left|X(z_0 + h) - X(z_0)\right|}{|h|^\alpha} = 0\right\}. \tag{12.1}$$

It measures the regularity strength of a signal at any given point. Geometrically speaking, this definition implies that the increments $X(z) - X(z_0)$ in the vicinity of z_0 are enclosed by a Hölderian envelope defined by $|X(z) - X(z_0)|^{\alpha_X(z_0)}$. A larger (smaller) value of $\alpha_X(z_0)$ corresponds to a smoother (rougher) signal at z_0.

12.2.2. Multifractional Brownian Motion

Fractional Brownian motion (fBm) is one of the most well-known stochastic fractal models. Initially introduced by *Kolmogorov* [1940] and developed by *Mandelbrot and Van Ness* [1968], it is defined as a zero-mean Gaussian process $B_H = \{B_H(z), z > 0\}$ characterized by a slowly decaying autocorrelation function depending on the Hurst parameter $0 < H < 1$:

$$E\left[B_H(z).B_H(s)\right] = \frac{\sigma^2}{2}\left[|z|^{2H} + |s|^{2H} + |z - s|^{2H}\right], \tag{12.2}$$

where $\sigma > 0$, $s > 0$, and $E[.]$ represents the expected value (for $H = 1/2$, the resulting process corresponds to the Brownian motion B (t)).

Since the fBm displays the same regularity everywhere, it cannot be used for investigating signals with a time/space-varying regularity. This limitation is avoided by introducing a local fractal model with an evolving H, specifically multifractional Brownian motion (mBm) [*Peltier and Lévy-Véhel, 1995; Benassi et al., 1997a*].

The mBm with the functional parameter $H(z)$ is the zero-mean Gaussian process defined as

$$W_{H(z)}(z) \propto \int_{-\infty}^{0} \left[(z-s)^{H(z)-1/2} - (-s)^{H(z)-1/2} \right] dB(s) + \int_{0}^{z} \left[(z-s)^{H(z)-1/2} \right] dB(s), \quad (12.3)$$

where $H : [0,\infty) \to (0,1)$ is a Hölder continuous function. In the case of a constant $H(z)$, $W_{H(z)}(z)$ is reduced to a simple fBm.

One most important feature of mBm is that its increments are generally neither independent nor stationary. In addition, in contrast with fBm, the point-wise Hölder exponent $\alpha_{W_H}(z)$ may vary with time/space and can be linked with the functional parameter H(z). Previous researches [*Peltier and Lévy-Véhel, 1995; Benassi et al., 1997b; Ayache and Taqqu, 2005*] demonstrate that for each z, $\alpha_{W_H}(z)$ and $H(z)$ are almost surely equal. The strength of the mBm process lies in its ability to model real phenomena characterized by punctual time (and/or space)-varying regularity.

12.2.3. Point-Wise Estimation of the Hölderian Regularity of mBm

The local Hölder function $H(z)$ at the point $z = i/(n-1)$ is given by [*Peltier and Lévy-Véhel, 1994*]:

$$\hat{H}(i) = -\frac{\log\left[\sqrt{\pi/2}\, S_{k,n}(i) \right]}{\log(n-1)}, \quad (12.4)$$

where $S_{k,n}(i)$ is the local growth of the increment process:

$$S_{k,n}(i) = \frac{m}{n-1} \sum_{j \in [i-k/2,\, i+k/2]} |X(j+1) - X(j)|, 1 < k < n \quad (12.5)$$

with n the signal length, k a fixed window size, and m the largest integer not exceeding n/k.

12.2.4. Hölderian Regularity-based Fine-to-coarse Reconstruction Algorithm

EMD assumes that well log data simultaneously possess many different modes of oscillations. Each mode corresponds to an IMF and can be generated from the data based on a local characteristic scale of the data themselves. An IMF must satisfy two conditions: (a) it has the same number of extreme and zero-crossings or differs by one at the most; (b) it is symmetric with the local zero value.

EMD can extract the IMFs collection via a sifting process, as follows:
1. Identify all the local extremes (maxima and minima) of the well log data $X(z)$.
2. Generate their upper $U(z)$ and lower $L(z)$ envelopes from, respectively, the local maxima and minima by cubic spline interpolation.
3. Calculate the point-by-point mean from the upper and lower envelopes: $m(z) = \dfrac{U(z) + L(z)}{2}$.
4. Extract the value $m(z)$ from the signal: $h_1(z) = X(z) - m(z)$.
5. Substitute the signal $X(z)$ by $h_1(z)$, and reiterate steps 1–4 until the obtained signal meets the two IMF conditions.

The sifting process is stopped by using one of the following conditions: after extracting n IMFs, the residue, $r_n(z)$, is either an IMF or a monotonic function. The stopping criteria are comprehensively discussed in earlier research [*Huang et al.*, 1998, 1999, 2003, 2008; *Huang*, 2005; *Flandrin and Gonçalvès*, 2004; *Rilling et al.*, 2003; *Rilling and Flandrin*, 2008].

At the end of this sifting procedure, the original signal can be expressed as

$$X(z) = \sum_{m=1}^{n-1} \text{IMF}_m(z) + r_n(z), \tag{12.6}$$

where $n - 1$ is the number of IMFs, i.e., the EMD of the signal consists of $(n-1)$ IMFs and one residual.

In this study, the result of the EMD is used to decompose the signal into a small-scale (high-frequency) component, a large-scale (low-frequency) component, and a trend component using the following Hölderian regularity-based fine-to-coarse reconstruction algorithm:

- For each $\text{IMF}_m(z)$ $(m = 1, n-1)$, compute the local regularity function $H_m(z)$ using the algorithm presented above, then calculate its average Hölder exponent value $\bar{H}_m$.
- The low-frequency component is constructed by summing IMF_m with $\bar{H}_m$ value greater than 0.5, while the partial reconstruction of the rest of IMF_m is identified as the high-frequency component. The residue $r_n(z)$ is designated as a trend component.

12.2.5. Multiple Layer Perceptron Artificial Neural Networks

Artificial neural networks (ANNs) have been used widely in engineering [*Chen et al.*, 1991; *Chen and Billings* 1992; *Calderón-Macías et al.*, 2000; *Castellanos et al.*, 2007]. This model represents an imitation of a biological neural network. The ANN is composed of neurons (processing elements, PEs), and connections between neurons (links), where each connection is associated with a weight. The multiple layer perceptron (MLP) ANN trained by a back-propagation algorithm is one of the most well-known and adaptable types of ANN and is successfully applied for nonlinear models. Its efficient performance depends on the selected training algorithm. It is shown that the Levenberg-Marquardt (L-M) algorithm yields better results than other training algorithms in terms of efficiency and high convergence speed [*Antcil and Lauzon*, 2004; *De Vos and Rientjes*, 2005; *Nayebi et al.*, 2006]. Here, the considered ANN architecture is a three-layer supervised learning model with an input layer consisting of explanatory variables (input parameters: log data), one hidden layer, and an output layer consisting of a single neuron representing the desired output (predicted *Vs log* data).

The network structure can be established in several ways, depending on the organization of the neurons. Each neuron acts by passing the result of summing its weighted input through a nonlinear activation function φ. A neuron *k* can be described by

$$u_k = \sum_{i=1}^{n} w_{ki} \, x_i \tag{12.7}$$

$$y_k = \varphi\left(u_k + b_k\right), \tag{12.8}$$

where $x_1, ..., x_n$ are the input signals, $w_{k1}, ..., w_{kn}$ are the synaptic weights of neuron k, u_k is the linear combiner output due to the input signals, b_k is the bias, and y_k is the output signal of the neuron. The error is then backward-propagated throughout the network to modify the weights of the connections and threshold, minimizing the sum of the mean squared error (MSE) in the output layer. The ANN calculates the weighted connection, minimizing the total MSE between the actual output of the ANN and the target output. The weights are adjusted using a momentum:

$$\begin{aligned} \left(w_{ki}\right)_{n+1} &= \left(w_{ki}\right)_n + \Delta w_{ki} \\ \Delta w_{ki}\left(n+1\right) &= \eta \delta_k x_i + \alpha \, \Delta w_{ki}\left(n\right), \end{aligned} \tag{12.9}$$

where η is a learning rate, δ_k is an error value for node k, α is a momentum term, n is the iteration number, and Δw_{ki} is the change of weight between two iterations. The momentum term is added to increase the speed of convergence [*Wang*, 2005].

12.3. Results and Discussion

As stated in our previous study [*Gaci*, 2014a], well logs are assumed to be paths of mBm processes. In addition, for a given well, the obtained regularity logs independently computed from well logs are strongly correlated. The Hölderian regularity is a significant attribute that can characterize subsurface heterogeneities.

Prior to implementing the MLP ANN algorithm, this statement is checked on *Vp* and *Vs* logs taken from four wells located in different settings: W1 (southwestern Algeria) and W2 (southeastern Algeria), HB (main KTB well, Germany), and VB (pilot KTB well, Germany). The sampling rate of all of these logs is 0.1524 m.

Using the algorithm presented above, the local regularity log is calculated for each well (Figures 12.1–12.4). Here, Hp and Vs refer to the regularity logs obtained from the Vp and Vs *logs*, respectively, for each well. The correlation coefficients between both regularity logs are 0.95, 0.80, 0.69, and 0.87 for wells W1, W2, KTB-HB and KTB-VB, respectively.

Therefore, the regularity profiles computed from P- and S-wave velocity logs are strongly correlated. This means that the regularity is a robust property of lithology and can thus be used as a supplementary input for neural network models to estimate well logs.

The suggested prediction methodology of Vs log from Vp log is based on the Hölderian regularity, EMD and MLP ANN algorithms. It is detailed as follows:

Step 1: Calculate the Hölderian regularity log from the Vp log using the algorithm developed above [*Peltier and Lévy-Véhel*, 1994].

Step 2: Use the EMD to decompose the Vp log data into a set of IMFs and one residue.

Step 3: Apply the Hölderian regularity-based fine-to-coarse reconstruction algorithm to extract a high-frequency (HF) component, a low-frequency (LF) component, and a trend component from the obtained IMFs and residue.

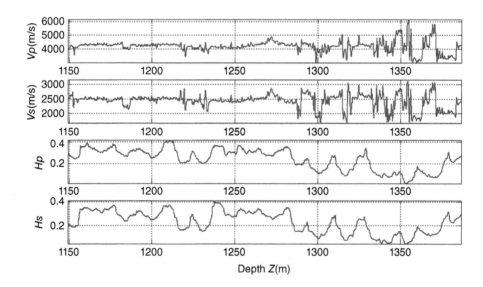

Figure 12.1 P- and S-seismic wave velocity logs (Vp and Vs) recorded at well W1, and their estimated regularity logs (Hp and Hs, respectively).

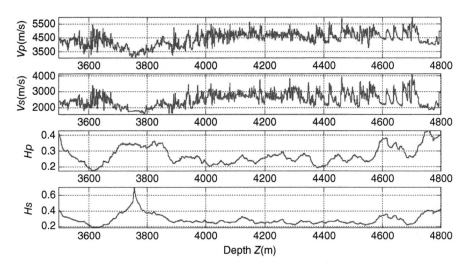

Figure 12.2 P- and S-seismic wave velocity logs (*Vp* and *Vs*) recorded at well W2, and their estimated regularity logs (*Hp* and *Hs*, respectively).

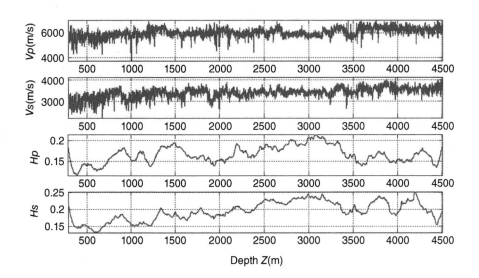

Figure 12.3 P- and S-seismic wave velocity logs (*Vp* and *Vs*) recorded at well KTB-HB, and their estimated regularity logs (*Hp* and *Hs*, respectively).

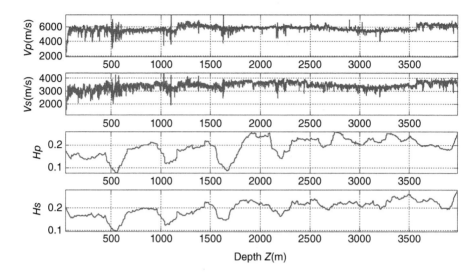

Figure 12.4 P- and S-seismic wave velocity logs (Vp and Vs) recorded at well KTB-VB, and their estimated regularity logs (Hp and Hs, respectively).

Step 4: Use the MLP ANN model to predict the Vs log at the desired well using the regularity *log* and the three reconstructed components as input parameters.

Figures 12.5 and 12.6 represent IMFs and residue that result from using EMD from the Vp and Vs logs, respectively, recorded in the KTB-HB well, while the regularity logs estimated from the IMFs corresponding to both logs are illustrated in Figures 12.7 and 12.8, respectively.

In addition, the LF, HF, and trend components are obtained from both logs using the Hölderian regularity-based fine-to-coarse reconstruction algorithm (Figures 12.9 and 12.10).

Here, four cases are examined to predict Vs logs from Vp logs using MLP ANN depending on the input parameters used:

Case 1: Vp log;

Case 2: Vp log and its associated regularity *log Hp*;

Case 3: Vp and the three components obtained from the EMD and the regularity-based fine-to-coarse reconstruction algorithms (HF, LF and trend)

Case 4: Vp, Hp and HF, LF and trend components;

The dataset used in this application contains the P- and S-wave sonics measured at the pilot and main boreholes (in short, VB and HB, respectively), drilled for the German Continental Deep Drilling Program (KTB).

The boreholes site is located in eastern Bavaria (Oberpfalz, Germany). The wells cross the crystalline metamorphic rocks of a Hercynian continental collision zone. Lithologically, this area is characterized by three main types of rocks: paragneisses, metabasites, and alternations of gneiss and amphibolites.

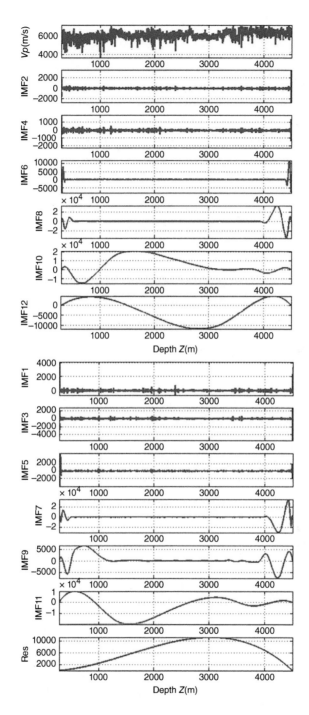

Figure 12.5 From top to bottom: Vp log recorded in well KTB-HB, 12 IMFs, and residue resulting from EMD (given in m/s).

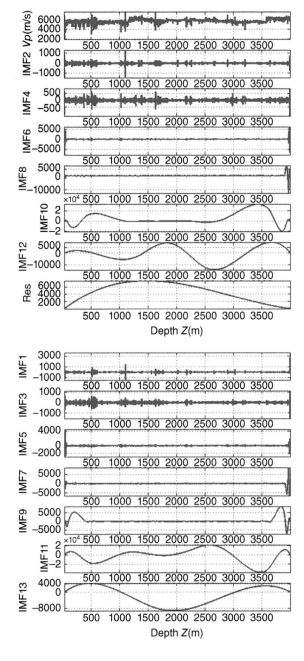

Figure 12.6 From top to bottom: *Vp* log recorded in well KTB-VB, 13 IMFs and residue resulting from EMD (given in *m/s*).

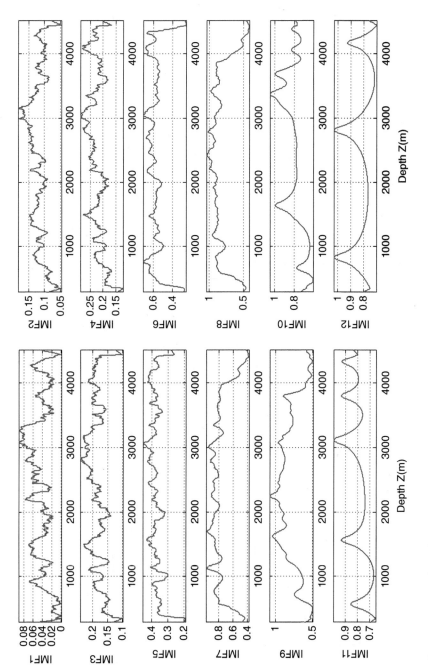

Figure 12.7 Regularity logs estimated from 12 IMFs of Figure 12.5.

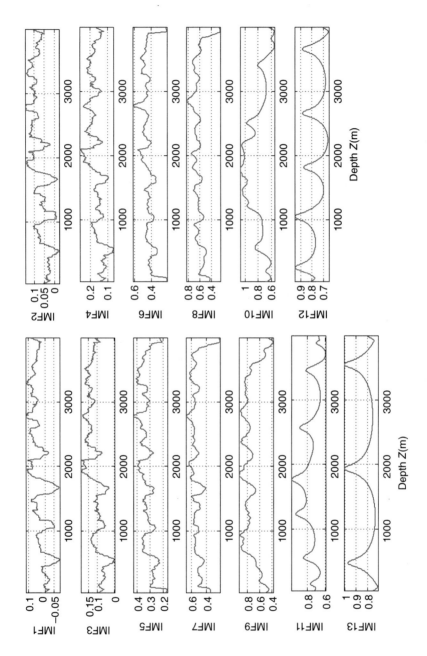

Figure 12.8 Regularity logs estimated from 13 IMFs of Figure 12.6.

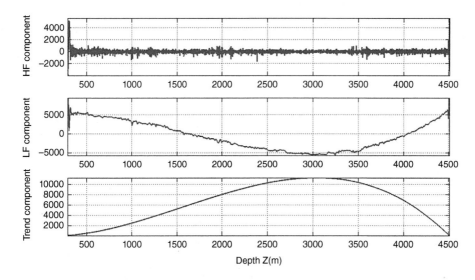

Figure 12.9 LF, HF, and trend components obtained from *Vp* log recorded at well KTB-HB using the Hölderian regularity-based fine-to-coarse reconstruction algorithm.

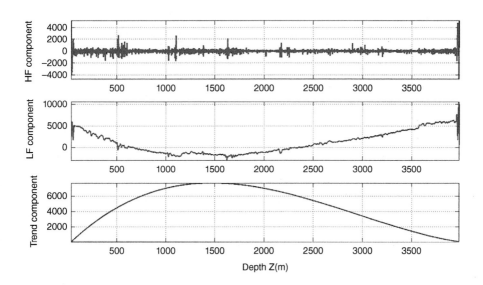

Figure 12.10 LF, HF, and trend components obtained from *Vp* log recorded at well KTB-VB using the Hölderian regularity-based fine-to-coarse reconstruction algorithm.

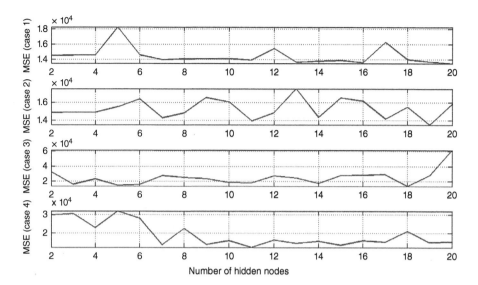

Figure 12.11 Training mean square error (MSE) as a function of the number of hidden nodes for the four considered cases.

The MLP ANN model is trained and tested on the main KTB borehole (KTB-HB) *log* data and then used to predict the *Vs* velocity log from the *Vp log* recorded at the pilot KTB borehole (KTB-VB). The training/test dataset model is made of 27,000 samples of *P*-wave velocity. Sixty percent of the data set was selected for the training set, while the remainder was used solely for the testing set and never considered for training the networks.

As previously mentioned, a three-layer supervised learning model is used. The outputs of the model consist of the predicted *Vs* values for different depths, while its input depends on the considered case. The number of hidden nodes was optimized by trial and error from 2 to 20. Figure 12.11 presents the MSE of the testing set as a function of the number of hidden nodes for the four studied cases. The optimized number of hidden nodes corresponds to the least MSE.

The results of the predicted *Vs log* from *Vp log* at the KTB-VB borehole for all the studied cases are presented in Figure 12.12.

For the sake of measuring the prediction performance, the MSE value is the main criterion considered to evaluate the prediction accuracy. The computed MSE values, expressed in (m/s)^2, corresponding to cases 1, 2, 3, and 4 are 73,765; 113,320; 63,773; and 56,749, respectively. As can be seen, the best prediction accuracy is obtained in case 4. For the purpose of completeness, *Vs* values are estimated using the formulae suggested by *Castagna et al.* [1985], *Han* [1986], and *Mavko et al.* [1998]. The MSE values obtained, given in (m/s)^2, are 6.6326×10^{10}, 3.7956×10^{10}, and 3.6536×10^{10}. The MLP ANN prediction using the

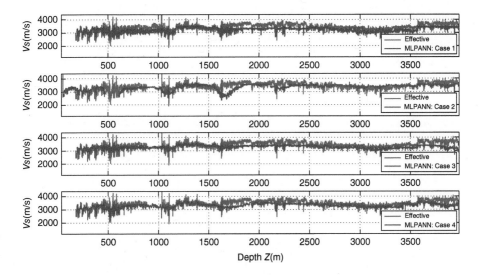

Figure 12.12 Estimations of S-seismic wave velocity (*Vs*) log at well KTB-VB using the dataset related to well KTB-HB corresponding to the four considered cases. In blue: MLP ANN predicted *Vs*; in red: effective *Vs* log recorded at well KTB-VB.

four-input selection outperforms the other techniques discussed in this study. However, it is worth noting that the predicted and measured Vs do not match in couple of zones. The potential reason behind those mismatches is the presence of major faults crossing the considered depth intervals.

12.4. Conclusions

This study presents the predicted *Vs log* data obtained from *Vp log* using the MLP ANN algorithm with different inputs: (i) only *Vp log*, (ii) *Vp* + its regularity *log Hp*; (iii) *Vp log* + three components: a high-frequency component, a low-frequency component, and a trend component, obtained by using the EMD and regularity-based fine-to-coarse reconstruction algorithms, and finally, (iv) *Vp* + *Hp* + HF + LF + trend. It is shown that the last selection leads to the most accurate predicted *Vs* log data with the lowest MSE value.

In fact, in the input parameters of MLP ANN the latter considers the Hölderian regularity of the velocity log, which is a robust property for characterizing lithology, and the different frequency components extracted using the EMD and the regularity-based fine-to-coarse reconstruction algorithms. It is more precise than the other techniques discussed. Therefore, it is very appropriate for prediction with nonlinear, nonstationary, and highly complex data, and is a very promising approach for *Vs log* forecasting.

References

Antcil, F., and N. Lauzon (2004), Generalization for neural networks through data sampling and training procedure, with applications to stream flow predictions. *Hydrology and Earth System Science*, 8(5), 940–958.

Ayache, A., and M. Taqqu (2005), Multifractional process with random exponent. *Publ. Mat.*, 49, 459–486.

Bekara, M., and M. Van der Baan (2009), Random and coherent noise attenuation by empirical mode decomposition. *Geophysics*, 74(5), V89–V98, doi:10.1190/1.3157244.

Benassi, A., S. Jaffard, and D. Roux (1997a), Elliptic Gaussian random processes. *Revista Matemática Iberoamericana*, 13(1), 19–90.

Benassi, A., S. Jaffard, and D. Roux (1997b), Gaussian processes and pseudo differential elliptic operators. *Rev. Mat. Iberoamericana*, 13(1), 19–81.

Bicego, M. and A. Trudda (2010), 2D shape classification using multifractional Brownian motion. *Lecture Notes in Computer Science*, 5342, 906–916.

Calderón-Macías, C., M. Sen and P. Stoffa (2000), Neural networks for parameter estimation in geophysics. *Geophysical Prospecting*, 48, 21–47.

Castagna, J., B. Batzle, and R. Eastwood (1985), Relationships between compression-wave and shear-wave velocities in elastic silicate rocks. *Geophysics*, 50, 571–581.

Castagna, J., M. Batzle, and T. Kan (1993), Rock physics: The link between rock properties and AVO response. In J. P. Castagna and M. Backus (eds.), *Offset-dependent reflectivity: Theory and practice of AVO analysis*, Investigations in Geophysics, 8 (pp. 135–171).

Castellanos, A., A. Martinez Blanco, and A. Palencia (2007), Application of radial basis neural networks for area forest. *International Journal Information Theories and Applications*, 14(3), 218–222.

Cersosimo, D., and J. Wanliss (2007), Initial studies of high latitude magnetic field data during different magnetosphere conditions. *Earth, Planets and Space*, 59(1), 39–43.

Chen, S., and S. Billings (1992), Neural networks for non-linear dynamic systems modeling and identification. *International Journal of Control*, 56, 319–346.

Chen, S., Cowan, C., and P. Grant (1991), Orthogonal least squares learning algorithm for radial basis function networks. *IEEE Transactions on Neural Networks*, 2(2), 302–309.

Chiew, F., M. Peel, G. Amirthanathan, et al. (2004), Identification of oscillations in historical global stream flow data using empirical decomposition. *Proc. 7th IAHS Scientific Assembly-Symposium on Regional Hydrological Impacts of Climate Variability and Change with an Emphasis on Less Developed Countries, Foz do Iguacu*, Brazil.

De Michelis, P., G. Consolini, and R. Tozzi (2012), On the multi-scale nature of large geomagnetic storms: An empirical mode decomposition analysis. *Nonlinear Processes in Geophysics*, 19, 667–673.

De Vos, N., and T. Rientjes (2005), Constraints of artificial neural networks for rainfall-runoff modeling: Trade-offs in hydrological state representation and model evaluation. *Hydrology and Earth System Science*, 7(5), 693–706.

Flandrin, P., and P. Gonçalvès (2004), Empirical mode decompositions as data driven wavelet-like expansions. *Int. J. Wavelets, Multires. Info. Proc.*, 2, 477–496.

Frayssinet, D., S. Thiria, F. Badran, et al. (2000), Use of neural networks in log's data processing: Prediction and rebuilding of lithological faces. *Petrophysics Meets Geophysics: Paris, France*, 6–8 November 2000.

Gaci, S. (2014a), On modeling oil borehole logs using local fractal processes: A case study from Algeria. In S. Gaci and O. Hachay (eds.), *Advances in Data, Methods, Models and Their Applications in Oil/Gas Exploration* (chap. 6, pp. 225–253), Science Publishing Group.

Gaci, S. (2014b), A multi-scale analysis of Algerian oil borehole logs using the empirical mode decomposition. In S. Gaci and O. Hachay (eds.), *Advances in Data, Methods, Models and Their Applications in Oil/Gas Exploration* (chap. 7, pp. 255–276), Science Publishing Group.

Gaci, S. (2014c), A Hilbert-Huang transform-based analysis of heterogeneities from borehole logs. In S. Gaci and O. Hachay (eds.), *Advances in Data, Methods, Models and Their Applications in Oil/Gas Exploration* (chap. 8, pp. 277–295), Science Publishing Group.

Gaci, S., and N. Zaourar (2010), A new approach for the investigation of the local regularity of borehole wire-line logs. *J. Hydrocarbon. Mines Environ. Res.*, 1(1), 6–13.

Gaci, S., and N. Zaourar (2011a), Heterogeneities characterization from velocity logs using multifractional Brownian motion. *Arab. J. Geosci*, 4, 535–541, doi:10.1007/s12517-010-0167-5.

Gaci, S., and N. Zaourar (2011b), Two-dimensional multifractional Brownian motion-based investigation of heterogeneities from a core image. In D. Chen (ed.), *Advances in Data, Methods, Models and Their Applications in Geosciences*, ISBN: 978-953-307-737-6, InTech. Available from: http://www.intechopen.com/articles/show/title/two-dimensional-multifractional-brownian-motion-based-investigation-of-heterogeneities-from-a-core-i.

Gaci, S., and N. Zaourar (2014a), A new regularity-based algorithm for analyzing Algerian airborne spectrometric measurements. *Proc. EGU 2014, Energy Procedia*, 59, 36–43.

Gaci, S., and N. Zaourar (2014b), On exploring heterogeneities from well logs using the empirical mode decomposition. *Proc. EGU 2014, Energy Procedia*, 59, 44–50.

Gaci, S., N. Zaourar, L. Briqueu, et al. (2011), Regularity analysis of airborne natural gamma ray data measured in the Hoggar area (Algeria). In D. Chen (ed.), *Advances in Data, Methods, Models and Their Applications in Geosciences*, ISBN: 978-953-307-737-6, InTech Available from: http://www.intechopen.com/articles/show/title/regularity-analysis-of-airborne-natural-gamma-ray-data-measured-in-the-hoggar-area-algeria.

Gaci, S., N. Zaourar, M. Hamoudi, et al. (2010), Local regularity analysis of strata heterogeneities from sonic logs. *Nonlinear Processes in Geophysics*, 17, 455–466. Available from: http://www.nonlin-processes-geophys.net/17/455/2010/.

Gloersen, P., and N. Huang, N. (2003), Comparison of inter annual intrinsic modes in hemispheric sea ice covers and others geophysical parameters. *IEEE Trans. Geosciences and Remote Sensing*, 41(5), 1062–1074.

Goutorbe, B., F. Lucazeau, and A. Bonneville (2006), Using neural networks to predict thermal conductivity from geophysical well logs. *Geophysical. J. Int.*, 166, 115–125.

Han, D. (1986), *Effects of porosity and clay content on acoustic properties of sandstones and unconsolidated sediments*. Ph.D. thesis, Stanford University.

Han, J., and M. Van der Baan (2011), Empirical mode decomposition and robust seismic attribute analysis. *CSPG CSEG CWLS Convention*, 114.

Hardy, H., and R. Beyer (1994), *Fractals in reservoir engineering*. Singapore, World Scientific.

Hewett, T. (1986), Fractal distributions of reservoir heterogeneity and their influence on fluid transport. *SPE Ann. Tech. Conf., New Orleans, Society of Petroleum Engineers (SPE)*, Paper 15386.

Huang, N. (2005), Introduction to the Hilbert-Huang transform and its related mathematical problems. In N. Huang and S. Shen (eds.), *Hilbert-Huang Transform and Its Applications*. World Scientific (pp. 1–26).

Huang, N., F. Schmitt, Z. Lu et al. (2008), Amplitude-frequency study of turbulent scaling intermittency using Hilbert spectral analysis. *Europhys. Lett.*, 84, 40010.

Huang, N., Z. Shen, and R. Long (1999), A new view of nonlinear water waves: The Hilbert spectrum. *Ann. Rev. Fluid Mech.*, 31, 417–457.

Huang, N., Z. Shen, S. Long, et al. (1998), The empirical mode decomposition method and the Hilbert spectrum for non-stationary time series analysis. *Proc. Roy. Soc. London*, 454A, 903–995.

Huang, N., M. Wu, S. Long, et al. (2003), A confidence limit for the empirical mode decomposition and Hilbert spectral analysis. *Proc. Roy. Soc. London*, 459A(2037), 2317–2345.

Huang, N., and Z. Wu (2008), A review on Hilbert-Huang transform method and its applications to geophysical studies. *Rev. Geophysics*, 46, RG2006, doi:10.1029/2007RG000228.

Kolmogorov, A. (1940), Wienersche Spiralen und einige andere interessante Kurven in Hilbertchen Raume. *Doklady*, 26, 115–118.

Li, M., S. Lim, B. Hu, et al. (2007), Towards describing multi-fractality of traffic using local Hurst function. *Lecture Notes in Computer Science*, 4488, 1012–1020.

Mandelbrot, B. (1975), *Les Objets Fractals: Forme, Hasard et Dimension*. Flammarion, Paris.

Mandelbrot, B. (1977), *Fractals: Form, Chance, and Dimension.*, W.H. Freeman and Co., San Francisco.

Mandelbrot, B., and J. Van Ness (1968), Fractional Brownian motion, fractional noises and applications. *SIAM Rev.*, 10, 422–437.

Mavko, G., T. Mukerji, and J. Dvorkin (1998), *Rock Physics Handbook. Tools for Seismic Analysis in Porous Media*. Cambridge University Press.

Nayebi, M., D. Khalili, S. Amin, et al. (2006), Daily stream flow prediction capability of artificial neural networks as influenced by minimum air temperature data. *Biosystems Engineering*, 95(4), 557–567.

Peltier, R., and J. Lévy-Véhel (1994), *A New Method for Estimating the Parameter of Fractional Brownian Motion*. INRIA, RR, 2396.

Peltier, R., and J. Lévy-Véhel (1995), *Multifractional Brownian Motion, Definition and Preliminary Results*. INRIA, RR, 2645.

Pickett, G. (1963), Acoustic character logs and their applications in formation evaluation. *Journal of Petroleum Technologies*, 15, 650–667.

Rilling, G., and P. Flandrin (2008), One or two frequencies? The empirical mode decomposition answers. *IEEE Trans. Signal Process.* 56(1), 85–95.

Rilling, G., P. Flandrin, and P. Gonçalvès (2003), Empirical mode decomposition and its algorithms. *IEEE-EURASIP Workshop on Nonlinear Signal and Image Processing*.

Wang, K. (2005), *Applied Computational Intelligence (CI) in Intelligent Manufacturing Systems (IMS)*. Advanced Knowledge International, Australia.

Wanliss, J. (2005), Fractal properties of SYM-H during quiet and active times. *Journal of Geophysical Research*, 110, A03202, 12. doi:10.1029/2004JA010544.

Wanliss, J., D. and Cersosimo (2006), Scaling properties of high latitude magnetic field data during different magnetosphere conditions. *Proc. 8th International Conference on Substorms, Banff, Canada*, pp. 325–329.

Williams, D. (1990), Acoustic log hydrocarbon indicator. *Society of Petrophysicists and Well Log Analysts, 31st Logging Symposium, Paper W*.

13

13. GEOPHYSICAL METHOD OF DEFINING RESIDUAL AND ACTIVE ROCK STRESSES

Kushbakali Tazhibaev and Daniyar Tazhibaev

Abstract

To define and examine the variation of stress in rocks created by mining minerals, including gas and oil, the geophysical method, a method based on the law of variation of transverse waves velocity and influenced by mechanical stresses, is developed. Examples of operating and residual stress definition of rocks are presented.

13.1. Introduction

At oil and gas exploitation, underground blowholes are formed, which leads to rock deformation. The stressed state in the earth's crust is modified. Rock deformation influences the output of oil and gas recovery. Through the development of minerals, including oil and gas, high variations of stress occur, especially nonequilibrium residual stresses in zones of their high concentration, leading to such catastrophic events as rock bursts and technogenic earthquakes. The definition of stresses is thus required for effective mining operations.

From experimental research [*Guz et al.*, 1974], it was shown that the velocity difference of the ultrasonic shift oscillations polarized in two main planes varies linearly up to the yield point of the material. In this context, the authors of that work consider that such measurements can be taken as a foundation of the stresses definition method.

Guz et al. [1977] noticed that with increasing dilatation stresses, the velocity of the wave polarized along the stresses decreases, but the velocity of the wave

Institute of Geomechanics and Mineral Development of National Academy of Sciences of the Kirghiz Republic, Bishkek, Kyrgyzstan

Oil and Gas Exploration: Methods and Application, Monograph Number 72,
First Edition. Edited by Said Gaci and Olga Hachay.
© 2017 American Geophysical Union. Published 2017 by John Wiley & Sons, Inc.

polarized across the stresses increases. By the action of compressing stresses, these dependences have the reverse character. With increasing compression, the velocity of the wave polarized along the stresses *increases*, and the velocity of the wave polarized across the stresses *decreases*. This characteristic of velocity changes for the polarized shift waves, according to *Guz et al.* [1977], allows us to define the values of stresses by measuring these velocities. *Guz et al.* [1977] noticed that the laws of the transition velocity of small variations through the characteristic base in materials with initial residual stresses (deformations) define the basic relations that are used to define the values and signs of stresses. At the same time, *Guz et al.* [1977] noticed that the absence of a uniform procedure for the definition of elastic constants of the second and third order leads to a wide range of received values of these constants, which are included in the calculation formulas for the definition of residual stresses, and thus reduce the accuracy of the results.

The work of *Gusha* [1983] considered the state and prospects of the ultrasonic method of defining the residual stresses and the acoustic interrelations of stresses for elastic waves in linear approximation. According to *Gusha* [1983], the resulting interrelations allow us to solve the inverse problem, using the measured values of elastic waves velocities propagation in the solid medium that define acting stresses into it. According to *Gusha* [1983], it is found that by defining the residual stresses we cannot always obtain the values of the initial velocity of the wave, corresponding to a nonloaded object (to a sample without residual stresses). Furthermore, in practice it is necessary to directly measure not the velocity but a value proportional to the velocity, for example, time, which is defined by not only active stresses but also the length of a wave path (the thickness of the object). This initial time (initial velocity) can be defined theoretically from the acoustic interrelations resulting in our work.

13.2. Problem and Solution Methods

The problem consists in defining the sign and magnitude of the rock stresses in space and time. Using the above-stated method [*Gusha*, 1983] for solution of that problem, it is necessary to apply analytical solution for stress definition. In our method, because of the requirement of complex performances, the definition, and the use of the theoretical solution results, the accuracy of the stress definition becomes essentially less.

There are limitations to the known mechanical methods of stress definition by the formation of new inner surfaces (unloading method). These are low accuracy because of incomplete changes in the existing residual stresses, depending on the dimensions and locations of newly formed surfaces (notches, holes); partial destruction of the structure; low informativity; and distortions of the initial stresses. The main limitations of the above-stated ultrasonic methods and the method of stress definition are their high labor requirement and low accuracy, because of the necessity to provide definitions of a considerable number of hard-to-define characteristics, after first carrying out additional research on the mechanical properties of stressed and nonstressed samples of the material. In the nonstressed samples, there may be residual stresses that distort the results and require additional definition of their value and sign, which is not provided by the specified methods.

As a result of determining the relative value of the polarized transversal wave's velocity from the investigation of stresses, we obtained formulae allowing us to define the sign and value of the active and residual stresses in solid materials in certain directions for a certain basis of measurement [*Tazhibaev*, 2011]. The formulae are as follows (Kushbakali Law) [*Tazhibaev et al.*, 2010, 2013]:

$$\sigma_X = \left(\frac{V_{SOZ}}{V_{SZ}} - 1\right) K_Z; \quad \sigma_Y = \left(\frac{V_{SOX}}{V_{SX}} - 1\right) K_X; \quad \sigma_Z = \left(\frac{V_{SOY}}{V_{SY}} - 1\right) K_Y; \quad (13.1)$$

where V_{SX}, V_{SY}, V_{SZ} are velocities of the cross-polarized ultrasonic wave passing through the directions X, Y, Z correspondingly; V_{SOX}, V_{SOY}, V_{SOZ} are velocities of the cross-polarized ultrasonic wave passing through the directions X, Y, Z correspondingly in the absence of stress (nonloaded state, without residual stresses); and K is the stress wave module of the material (the name we have given this), whose dimension is stress.

Let us introduce such designations:

$$\omega_Z = \left(\frac{V_{SOZ}}{V_{SZ}} - 1\right); \quad \omega_X = \left(\frac{V_{SOX}}{V_{SX}} - 1\right); \quad \omega_Y = \left(\frac{V_{SOY}}{V_{SY}} - 1\right).$$

Substituting them into formula (1), we obtain

$$\sigma_X = \omega_Z K_Z; \quad \sigma_Y = \omega_X K_X; \quad \sigma_Z = \omega_Y K_Y,$$

and

$$K_X = \frac{\sigma_Y}{\omega_X}; \quad K_Y = \frac{\sigma_Z}{\omega_Y}; \quad K_Z = \frac{\sigma_X}{\omega_Z} \quad (13.2)$$

for isotropic materials $K_X = K_Y = K_Z = K$. For the anisotropic (layered, crystal, anisotropic) materials, the value of the stress wave module K is defined in the corresponding directions. The value K is defined from sounding experiments at loading and unloading of the samples. We perform 5 to 10 definitions of value K at loading and unloading of the representative volume sample of the investigated material. The value K as the characteristic of the material is defined as an average value from 10 to 20 individual values obtained by ultrasound during the process of loading and unloading of the material sample. It should be noted that the value of K must be defined for the characteristic that is representative volume, because that module depends on the structure and material content of the specimen. In this connection, to define the value of K we next offer the dimensions of the prismatic sample as $5 \times 5 \times 10$ cm and $7 \times 7 \times 14$ cm. Experimental definition of the stress wave module of the material (K) is provided in the following sequence and on the device in Figure 13.1.

1. On the prism side, with the dimensions specified above, in its middle part the transmitter and the receiver of the shear wave is located, combining their polarization vectors among themselves and with the direction of compressing stress (σ_z).
2. The velocity of the ultrasonic polarized shear wave in the absence of the load (stress) – V_{SOY} is measured.
3. The hydraulic press loads the prism stepwise, and at each step of uniaxial compression (on each step we added 1000 or 2000 kg) the velocity of the polarized shear wave – V_{SY} is then measured.
4. For each step of the load, the value of stress is defined by the ratio of the value of load (according to the load measure of the hydraulic press) to the cross-sectional area of the prism.
5. Using formula (2), the values of the stress wave module K are defined and its average value is taken for further calculations. Measurements of the ultrasonic polarized shear wave velocity and the definition of the values of K can be done also by unloading the specimen.

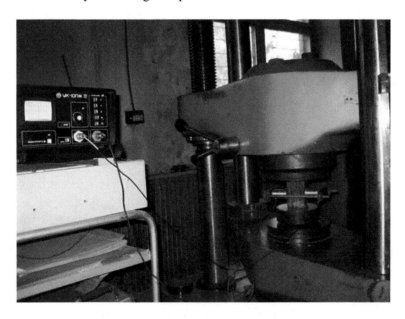

Figure 13.1 The device for measuring stresses.

With the defined value of the stress wave module, having measured the velocities of the shear polarized ultrasonic waves, it is possible to define the residual and active stresses using formula (1). It should be noted that active stresses can also include the residual stresses (when present, for example, in rocks). In this connection, the residual stresses, if necessary, are defined separately, free from the external loading of the representative pieces of rock taken from the location of the in situ velocity wave measurements of rocks. Usually, residual stresses are defined under laboratory conditions.

The active stress in the rocks is defined using the following procedure:

1. Two smooth surfaces are created in the rocks parallel to the set axis, e.g., to axis Z (vertical direction) with a distance between them of about 7 cm (the base) forming gaps for location of shear waves transducers.
2. The surfaces are refined and dried, and then we place a shallow layer of a contact medium (polysaccharides, artificial honey) on the smooth surfaces, together with the transducers, and press the transducers lightly (with a constant force) onto the smooth surface of the rock, with their polarization vectors directed on the set axis (Z).
3. The transducers are connected to the ultrasonic device ($yk - 10\,PM$) and, after heating the device for 20 minutes not less than 10 times, we measure the time of the shear wave passing through the set base (7 cm) in an automatic mode and with the accuracy of time measurement ± 0.01 mks; then, using the average value of measured time, we define the velocity speed of the polarized shear ultrasonic wave.
4. For the stress monitoring procedure (in the case of essential change of stress owing to fast movement of the mining face, which influences the stress state), the device and transducers are left in the mine for a control period and the measurements of the time of shear waves passing are periodically made according to point 3.
5. In the case of defining active stresses after measuring the time of shear waves passing according to point 3, the transducers are removed, and a piece of rock with dimensions $7 \times 7 \times 15$ cm is cut out from the same place where the shear wave passing time was measured, keeping the base (7 cm).
6. From the selected piece, a prism with dimensions $7 \times 7 \times 14$ cm is produced. For this prism, the stress wave module K is defined based on the sequence specified above, passing the polarized shift wave at different levels of the compression load on the base (7 cm), on which the research was conducted in situ in the rocks.
7. Using the velocity of the polarized shear wave measured in situ in the rocks, and the value of the stress wave module K, and also the velocity of the polarized shear wave for the nonloaded state, the vertical (for example) component of normal stress of the rocks using formula (1) is defined. The stresses in other directions are defined in the same manner.

For the definition of residual stresses, in the cubic sample with the dimensions $7 \times 7 \times 7$ cm, free from external loading, it is necessary to measure the velocity of the passing ultrasonic polarized shear wave in the following sequence.

1. Using ultrasonic soundings in directions through every angle via 1° or 10° from zero to 180°, by turning the transmitter and the receiver of the acoustic polar scope [*Gorbatsevich*, 1985], we can define the values of the velocity of the shear polarized wave passing V_{SI} for parallel and perpendicular (to a considered direction) vectors of polarization for each direction on three orthogonal planes.

2. On the basis of the equality of time or velocity of passing parallel V_{SP} and perpendicular V_{SC} of the shear polarized waves for the set base of soundings, the velocity of the shear wave in the case of the absence of residual stresses is V_{SOI} (one of the signs of absence of residual stresses is defined by the equality $V_{SP} = V_{SC}$).

3. In the case of not detecting a direction or location without residual stresses, we have to fix all the values of the velocity of the ultrasonic polarized shear wave passing in all the above-stated directions and on three orthogonal planes, and then the residual stresses are completely extracted by a known method of the researched sample.

4. The velocity of the ultrasonic shear polarized wave for the sample without residual stresses (after the removal of residual stress), V_{SOP} and the average value of the stress wave module K in the above-stated sequences are defined.

5. Using the values of the velocities of waves for different directions and planes, and also the values of the velocity for the nonloaded state V_{SOI} and the average value of the stress wave module, we can define the values of residual stresses according to formula (1).

13.3. Results

By passing ultrasonic soundings in directions Z, Y, X, namely the ultrasonic polarized shear waves, it is possible to define the normal stresses orthogonal to the specified directions, and the main normal stresses on the corresponding planes, by turning the transmitter and the receiver of the polarized shear wave through every 1° or 10° from zero to 180°, synchronously turning the transmitter and receiver of the acoustic polar scope. After defining the main normal (maximum and minimum) stresses, it is possible to define also the tangential stresses.

Figure 13.2 presents an example of the definition of the results for residual stresses in the sample (dimensions $11.2 \times 7.27 \times 3.43$ cm, base 7.27 cm) of metasomatic deposits at Kumtor.

The results of the values of the actual stress σ_Z, defined by direct experimental measurements of the loading device, are compared with values received under formula (1). For fine-grained marble (Toktogul), the ultrasonic values of

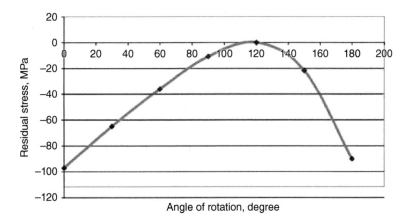

Figure 13.2 Values of residual stresses of the metasomatic sample (from one area, Kumtor mine, hallmark RS No. 5, sample 3–4).

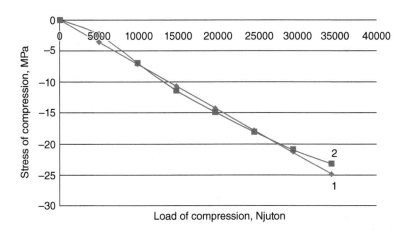

Figure 13.3 Stress components σ_z (1: on a force counter, 2: using formula (1), loading, fine-grained marble, Toktogul, sample 1–5).

stress obtained using formula (1) are in good agreement with the values of stresses obtained by direct measurement on a compression force counter (Figure 13.3).

13.4. Discussion of Results

It should be noted that the values of the stresses defined by formula (1) represent the average value of stress for the set bases. Further research in that direction is necessary in anisotropic, layered, and cracked rocks. In layered and cracked rock, the stress wave module *K* depends on the material structure. For example, the value of that module along layers of the rock will be greater than across layers, and the stress in such anisotropic rocks can be defined by taking into account their frame and texture.

13.5. Conclusions

The proposed method based on the proposed law of velocity change of the polarized transversal wave from stresses allows us to define and monitor the active stresses in rock pillars and the walls of mountain mines in quasi-homogeneous rocks and also in solid constructions consisting of homogeneous technical materials. Thus, it is necessary to solve the technical problems of forming smooth surfaces in the rock pillars and walls of mountain mines to install transmitters and receivers of waves, for definition and monitoring of the components of stresses in the given directions. As the formula for the definition of stresses includes relative values of velocity (time) of passing of the polarized transverse waves, and the basis of measurements in laboratory and full-scale experiments do not differ, the measurement in situ of stresses in rocks by using this method is obviously possible once all the required technical problems are solved.

References

Gorbatsevich, F. (1985), Acoustopolarization measurements of characteristics of anisotropy of rocks. Apatite, Kola Centre of Science of the Russian Academy of Sciences (in Russian).

Gusha, O. (1983), Ultrasonic method of definition of residual stresses, states and prospects. *The Ex-Lane Methods of Researches of Residual Stresses*, Kiev, pp. 77–81.

Guz, A. et al. (1974), Bases of an ultrasonic non-destructive method of stresses definition in solids. Kiev, Naukova Dumka (in Russian).

Guz, A. et al. (1977), Foundations in acoustoelasticity. Naukova Dumka, Kiev (in Russian).

Tazhibaev, K. (2011), Law of change of speed of transit of the polarized cross-ultrasonic wave from stress in firm materials and its application. *Bulletin of the Kyrgyzsko-Russian Slavic University*, 11(11), 151–156 (in Russian).

Tazhibaev, K. et al. (2010), Means of definition of residual stresses in firm materials. *Patent 1245 KP, from 26 February* (in Russian).

Tazhibaev, K. et al. (2013), Law of change of passing velocity relative value of the ultrasonic polarized shear wave from mechanical stress in solid materials (Kushbakali's Law). *Diploma 453 on a discovery from 3 October 2013. Moscow.* International Academy of Authors of Discoveries and Inventions, Russian Academy of Natural Sciences (in Russian).

14

14. ON THE POSSIBILITY OF USING MOBILE AND DIRECT-PROSPECTING GEOPHYSICAL TECHNOLOGIES TO ASSESS THE PROSPECTS OF OIL-GAS CONTENT IN DEEP HORIZONS

Sergey Levashov[1,2], Nikolay Yakymchuk[1,2], and Ignat Korchagin[3]

Abstract

The results of evaluation of the prospects of a number of oil-bearing areas and structures in the Dnieper-Donets Basin and the Caspian Basin are analyzed. These were obtained with the mobile technology of frequency-resonance processing and by interpretation of remote sensing data used for the "direct" prospecting and exploration of ore mineral and fossil fuels. We can say, in general, that the interpretation of the remote sensing data yields new (additional) and mainly independent information that can be useful and in demand at any stage of oil and gas prospecting. An important characteristic feature of this information is the prompt time of receiving it. Applying the method of the maximum value of fluid pressure in assessing reservoirs allows us to narrow even more significantly the priority areas for conducting detailed exploration and locating wells for exploration.

[1]Institute of Applied Problems of Ecology, Geophysics and Geochemistry, Kiev, Ukraine
[2]Management and Marketing Center of Institute of Geological Science NAS Ukraine, Kiev, Ukraine
[3]Institute of Geophysics of Ukraine National Academy of Science, Kiev, Ukraine

Oil and Gas Exploration: Methods and Application, Monograph Number 72,
First Edition. Edited by Said Gaci and Olga Hachay.
© 2017 American Geophysical Union. Published 2017 by John Wiley & Sons, Inc.

14.1. Introduction

When developing the hydrocarbon potential of deep cross-section horizons, the material (financial) costs for drilling wells are significantly increased. This requires additional assessment of the recommendations made for drilling. As one way to receive additional and independent information in a more timely manner, mobile geophysical methods can be used when conducting prospecting and exploration for oil and gas accumulations.

Since 2010, the authors have been testing the mobile technology of frequency-resonance processing and interpretation (decoding) of remote sensing data for combustible and ore minerals, and "direct" searching for and prospecting of aquifer reservoirs [*Levashov et al.*, 2010, 2011]. This technology allows us to detect and map anomalous zones of the "oil reservoir," "gas accumulation," "hydrate reservoir," and "water-bearing reservoir" type, etc. To date, the technology has been tested on more than 150 individual sites and areas. Its integrated use with mobile geoelectrical methods of forming a short-pulsed electromagnetic field (FSPEF) and vertical electric-resonance sounding (VERS) [*Levashov et al.*, 2008, 2012c, *Yakymchuk et al.*, 2008] can significantly reduce both the time required for fieldwork in remote and difficult areas to access (tundra, taiga, mountain areas, shallow shelf, etc.), and its cost. In papers by *Levashov et al.* [2012a, b], there are examples of the application of these mobile methods [*Levashov et al.*, 2008, 2010, 2011, 2012c] to evaluate the hydrocarbon potential of the deep cross-section horizons. Below, we analyze the results of applying the frequency-resonance technology of remote sensing data processing for the hydrocarbon (oil and gas) potential of deep horizons (including presalt deposits) assessed within selected areas in the Dnieper-Donets (DDB) and Caspian Basins.

14.2. Mobile Geophysical Technology

The practical application of technology in the exploration process for various minerals allows us to significantly speed up the search process itself, as well as to increase its efficiency.

14.2.1. Technology Components and Equipment

Mobile technology of geophysical research includes the following specific prospecting methods, instrumentation, and software:

a. The frequency-resonance method of remote sensing (satellite) data processing and interpretation [*Levashov et al.*, 2010, 2011, 2012c; *Solovyov et al.*, 2011]
b. Areal survey by geoelectrical method of FSPEF [*Levashov et al.*, 2008, 2012c, *Yakymchuk et al.*, 2008]
c. A method of VERS [*Levashov et al.*, 2008, 2012c, *Yakymchuk et al.*, 2008]

d. Computerized equipment for the field observations; GPS receiver; software for acquisition, processing, and interpretation of the measurement data; and the technique of field observations.

14.2.2. Problem Solving

In the process of testing and practical application, the mobile technology of geophysical research has been used repeatedly for the following problem solving:

a. Identification and mapping in cross-section of the deposit-type anomalies (DTAs), which may be due to hydrocarbons accumulations, ore minerals, or water-bearing reservoirs;

b. Determining the depth and thickness of anomalous polarized layers (APLs) of the oil, gas, ore-bearing stratum, and aquifer type;

c. Performing in a short time reconnaissance surveys of large and difficult to access territories with possible oil and gas and ore minerals;

d. Performing detailed works within individual anomalous zones and selecting promising objects for well locations, making predictive assessments of hydro-carbon reserves and ore minerals, making decisions about the directions of further geological and geophysical surveys and drilling;

e. Detection and mapping of areas of high gas saturation in the coal bed and surrounding rocks within mine fields;

f. Mapping fracture zones and hydrocarbon accumulations in fractured parts of the crystalline basement;

g. Exploration for oil and gas carried out from vessels at sea, etc.

14.2.3. Testing and Efficiency

The mobile technology has been tested on more than 150 sites of ore, oil, and gas fields and in promising for various minerals areas [*Levashov*, 2012; *Levashov et al.*, 2008, 2010, 2011, 2012a, b, c]. Geography of approbation: Ukraine, Republic of Kazakhstan, Russia, Belarus, Turkmenistan, Syria, Colombia, United States, Mongolia, Turkey, Slovakia, Sudan, Tunisia; shelves of the Black, Azov, Barents, Caspian, Pechora and Mediterranean Seas; Gulf of Mexico; shelves of Sakhalin, Vietnam, Cambodia, Indonesia, Venezuela, Trinidad and Tobago, Antarctic Peninsula, Falkland Islands [*Levashov et al.*, 2008, 2010, 2011, 2012a, b, c, d, 2014, 2015, *Yakymchuk et al.*, 2015].

14.2.4. Stages of Work

Prospecting investigation by the mobile method of frequency-resonance processing of remote sensing data and ground-based geoelectric methods FSPEF and VERS may be conducted in two main phases.

Stage 1: Defining the petroleum potential (ore content, water content) of surveyed areas and sites from the results of processing and interpretation of remote sensing data.

Stage 2: Detailed research by ground-based geoelectrical methods, FSPEF and VERS, of the promising sites and anomalous zones defined.

14.2.5. The Features of Mobile Direct-Prospecting Methods

The ground-based geoelectrical methods of FSPEF and VERS make possible the efficient and accurate determination of a geologic model beneath a sounding site. The original FSPEF and VERS methods are based on studying the geoelectrical parameters of the medium in pulsed transient geoelectrical fields, as well as of the quasi-stationary electric field of the Earth and its spectral features over hydrocarbon reservoirs [*Levashov et al.*, 2008, 2012c, *Yakymchuk et al.*, 2008, 2014].

Areal survey by FSPEF method can be performed in two ways: on the car and by the pedestrian. In the automotive embodiment, the antennas are mounted outside the vehicle. The survey is conducted at a speed not more than 20–30 km per hour. The signals of field formation are registered every 50–60 meters or through a priori specified time interval.

In the pedestrian embodiment, the equipment, including the batteries, is transferred by the operators (weight of apparatus: up to 10 kg). The survey is performed by two operators.

The results of FSPEF survey are used for selecting optimal profiles for the VERS method, which are implemented at individual points. VERS profiles intersect anomalous zones mapped during the FSPEF survey. The time required for sounding at one point within the depth interval up to H = 1.0 km is about 2 hours. The sounding is performed by three operators.

The main difference between the used geoelectrical methods from classic electric prospecting is that in the electric prospecting methods the surface-to-air boundary is considered a conductor and an insulator. In geoelectrical method the near-surface part of the atmosphere is considered as weakly ionized plasma. The frequency generation occurs in the space charge of near-surface part of the atmosphere.

In the natural quasi-stationary electric field Ez, the anomalous object, located at a depth H, forms with the ground-air interface the polarized dipole of H thickness. During the natural or artificial changes (excitation) of the primary polarizing field Ez, the dipole radiates electromagnetic waves with frequencies $f = C/L$, where C is the speed of light and L is the length of the waves, which is equal to $L = 2H$.

Feature of measurement by VERS method is following. The natural field Ez is distorted over exploratory geological object by short electromagnetic pulses, and at this point the frequency responses are measured. When the frequency response is known, we can determine the depth of the object: $H = C/2f$.

The generator of rectangular impulse with the frequency of 3.0 kHz is used for the perturbing field generating. The registering of the natural electromagnetic field is carried out by using broadband antennas. Signals are recorded in the frequency range from hundreds of hertz to tens of megahertz. Spectral signal processing is carried out further; the results of processing are compared with the reference spectra of searching objects.

The distinctive features of the FSPEF and VERS geoelectrical methods, used by the authors, are described in articles by *Levashov et al.* [2012c] and *Yakymchuk et al.* [2008, 2014]. The principle of vertical sounding was done by *V. N. Shuman* [2012]. A technique similar to the VERS method of sounding is described in the patent from *Weaver and Warren* [2004]. For more information about the features of mobile technologies and the results of their application, visit the EAGE site, www.earthdoc.org/.

14.2.5.1. The Frequency-Resonance Processing of Remote Sensing Data

Currently, there are methods of remote sensing data processing and interpretation that are developed and improved within the framework of the "substantial" ("matter") paradigm of geological and geophysical studies. The essence of this paradigm is "direct" searching and prospecting for specific substances such as oil, gas, gold, silver, platinum, zinc, iron, water, etc. [*Levashov et al.*, 2012c]. Among such technologies may be Infoscan (http://www.infoscan.ru), Tomko [*Rostovtsev et al.*, 2011], Poisk [*Kovalev et al.*, 2009, *Pukhliy et al.*, 2010], etc. The effectiveness of geophysical methods, based on the principles of this paradigm, is higher than traditional.

A phenomenological description of features of the Tomko technology is the following: "It is based on the latest advances in astrophysics, mathematics, and knowledge about electromagnetic radiation, modern laser, computer hardware and software. It became possible due to the phenomenological approach to the study of the phenomena observed. Technology of quantum-optical filtering of satellite imagery allows, in most cases, anywhere in the world to identify the boundaries of projected oil and gas fields and record the density distribution of reserves within them" [*Rostovtsev et al.*, 2011, p. 61].

Theoretical justification for this technology may be the ideas of E. I. Tarnovskiy, which are based on the fact that all atoms in molecules have a certain spatial position and its own electro-magnetic field with the characteristic spatial frequency intensity distribution.

The spatial-frequency structure of electromagnetic fields of any substance is determined by the chemical composition and molecular structure, or spatial lattice material. A large number of homogeneous substances will create a collective characteristic for the substance electromagnetic field radiation, whose power is proportional to the concentration of a substance in a given direction. We may assume that the linear polarized wave with the specified frequency response

Table 14.1 Resonance frequencies of water of different salinity.

Number	Mineralized Water	Mineralization (g/dm³)	Resonance frequencies, kHz
1	Structured (alpine sources)	<0.1	717.6
2	Weakly mineralized (Morshynskiye sources)	0.1–0.4	643.8
3	Average mineralization	0.5–1.0	615.7
4	Strong mineralization	5.0–15.0	551.5

carries information about the structure of the substance, is not absorbed by the medium, and their intensity does not decrease with distance. In this case, a homogeneous substance at an arbitrary depth of the earth will create a field, as if this substance were on the surface.

It was found that the characteristic of an electromagnetic wave of large quantities of oil and gas is fixed in a certain way on the satellite image; this is already used for opening and identifying yet-unknown fields.

We pay attention to the phrases "a large number of homogeneous material," "radiation power which is proportional to the concentration of the substance," "will create a field, as if it were a substance on the surface" in the above quotation.

In the Tomko technology, the useful signal separation is conducted by the quantum-optical filtering of satellite images. In this method, used by the authors, the separation of the useful signal from the satellite images is provided by the frequency-resonance method. For the various minerals (oil, natural gas, uranium and gold, water, zinc, etc.), their characteristic resonance frequencies have been identified on the minerals samples, and these are used during the remote sensing data processing and decoding.

Table 14.1 below gives the values of the resonance frequencies for water of different mineralizations. For other minerals, the resonance frequencies are fixed in a wider range.

For frequency-resonance processing, the multispectral images are used from different satellites. In most cases, these are in the public domain. For reconnaissance surveys of large areas and during the processing of remote sensing data on a scale of 1:50,000 and smaller, the images of Landsat 5 and Landsat 7 with a resolution of 30 m/pixel can be used. When looking for small objects, the remote sensing data processing should be done on a larger scale and requires high-resolution images, 2.5–1 m/pixel.

14.2.5.2. Assessment of the Values of Fluid Pressure in the Reservoir

Within the frequency-resonance technology of remote sensing data processing in general, a special place occupies a method of maximum values of fluid pressure assessment in the reservoir [*Levashov et al.*, 2011]. First, it allows researchers to significantly narrow the areas for hydrocarbon deposit searching and, consequently, the sites for the exploratory wells. Second, the result of its application for the fluid pressure assessment in reservoir can help to form an initial hypothesis about the depth of hydrocarbon deposits. Third, the lack within detected and mapped anomalous zones of the reservoir type of areas with relatively high values of fluid pressure in reservoir allows researchers to exclude such areas (anomalies) from the list of objects that deserve priority for a detailed study and drilling. Assessments of maximal values of fluid pressure in reservoir, in which the anomalous zones of oil (gas) reservoir type are constructed, are complex parameters that depend on the pressure of the gas in the fluid or in the free form in reservoir, as well as its amount, i.e., the rocks' porosity. Therefore, at the edges of the anomalous zones, a decrease of this parameter value is fixed. In areas of gas absence, the pressure is not defined, so here this parameter value is zero.

When research was conducted in 2015, the authors started testing the improved method of estimating the values of fluid pressure in the reservoirs. A distinctive feature of the improved method is that it allows to obtain the estimates of fluid pressures in the predicted hydrocarbon reservoirs within a priori given (in each case) intervals of cross-section (from the surface to a depth of 6 km, for example). In this situation, the process of anomalous response registration does not stop even in the case of their absence within certain intervals of the cross-section. The experiments showed that such a methodological procedure is justified; the anomalous responses within many detected anomalous zones were fixed at various intervals (segments) of the resonance frequencies. Therefore, an improved method of estimating fluid pressure allows researchers to detect the predicted hydrocarbon deposits in the various horizons of the cross-section and to assess approximately the depths of their occurrence.

The following results were obtained, in general, with the mobile technology of frequency-resonance processing and interpretation of remote sensing data using *Levashov et al.* [2010, 2011]. By publishing the material obtained using this mobile method, we are trying to draw the attention of scientists and experts from oil companies and geophysical service companies to the potential opportunities that such mobile technologies for "direct" prospecting offer, allowing us to obtain solutions to specific (concrete) problems at different stages of the exploration and development of oil and gas deposits.

14.3. Investigation Results in Dnieper-Donets Basin

14.3.1. Oil and Gas Prospect Area in the Poltava Region

In 2006, ground-based geoelectrical studies using the FSPEF and VERS methods [*Levashov et al.*, 2012a] were conducted in the search area. Here, DTAs with an area of over 20 km² were discovered and mapped by the FSPEF survey.

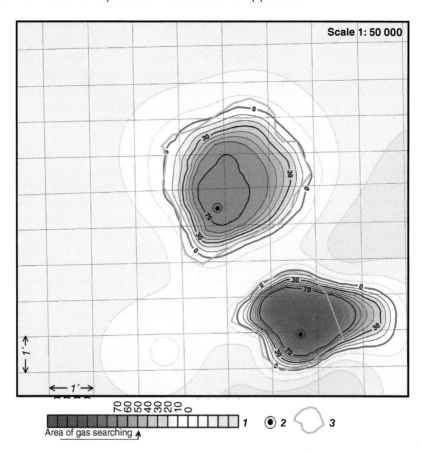

Figure 14.1 Map of anomalous zones of the gas deposit type in the oil and gas prospecting area in the Dnieper-Donets Basin (based on remote sensing data interpretation), Poltava region, 2011. 1 = scale of gas pressure values inside the reservoir, MPa; 2 = locations recommended for VERS sounding and drilling; 3 = anomalous contours from forming a short-pulsed electromagnetic field (FSPEF) survey data.

In the depth interval of 5200–5800 m, the presence of APLs of the gas and gas-condensate type were established by VERS. In 2011, the remote sensing data from this area were processed further (Figure 14.1).

As a result, within the mapped DTAs, it was possible to highlight regions with relatively high values of the maximum fluid pressure in the reservoir, which further confirms their prospects. In Figure 14.1, the anomalies contour line with a value of 70 MPa delineates an area where there may be inflows of hydrocarbons from the depth interval up to 7000 m.

14.3.2. Gas Field in Poltava Region

Processing and interpretation of satellite data over the area of a known gas field was conducted also using techniques of the maximum values of gas pressure in assessing the reservoirs [*Levashov et al.*, 2011].

As a result of the survey, a relatively large anomalous zone of "gas accumulation" type was highlighted and mapped within the area (Figure 14.2). A DTA was also found in the eastern area of remote sensing data processing, adjacent to its boundary. Third, a small fragment of the anomalous area was found on the southern border area of remote sensing data processing.

In Figure 14.2, the anomalous zones are given in isolines of the maximum values of fluid pressure in the reservoir (MPa). The contour line with the value of 54 MPa outlines the optimal site for gas exploration at the depth interval of 5200–5800 m (at the depth of the deposits in the B17 and B21 horizons location, initial reservoir pressures are 54.13 MPa and 58.63 MPa). The areas of mapped

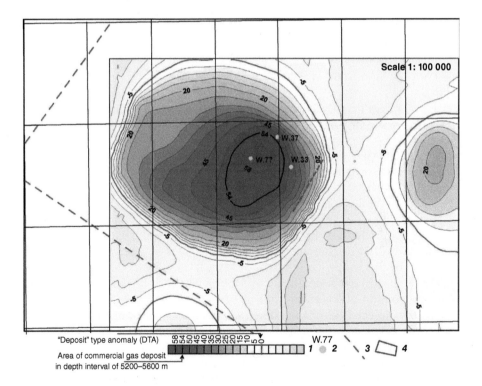

Figure 14.2 Anomalous zones of gas deposit type in the area of the gas field (based on interpretation of remote sensing data), Poltava region. 1 = scale of values of gas pressure in reservoir, MPa; 2 = drilled wells; 3 = tectonic faults; 4 = outlines of the licensed area.

anomalous zones are as follows: Central: total (zero contour line) $= 44.0\,km^2$, along the isoline of $54\,MPa = 4.3\,km^2$; Eastern: total $= 9.8\,km^2$. The maximum value of fluid pressure in the reservoir in the region of Eastern anomalies is about $20\,MPa$ (Figure 14.2). This indicates that commercial inflow of hydrocarbons from the interval of the B17 and B21 horizons location is unrealistic.

Production well 77 (Figure 14.2) is located in the center of the anomaly with a contour line of $58\,MPa$ (maximum value of fluid pressure in the reservoirs). Unproductive wells 33 and 37 are located outside the contour of 54. It is possible that in these wells industrial gas inflows were not obtained due to the relatively low values of the fluid pressure in the reservoir (compared to the hydrostatic pressure at the depths of the location of the reservoirs). In the unproductive wells, the deterioration of the reservoir properties of productive horizons was identified.

Detailed research within this field was also conducted by the FSPEF and VERS geoelectrical methods [*Levashov et al.*, 2008, 2012c]. The results of the survey using the FSPEF method are shown in Figure 14.3 and the VERS data at one point in Figure 14.4. These data indicate that the ground survey by the FSPEF method allows substantial localization of the areas of search, and of exploratory and production wells. VERS enables us to determine the bedding depths and thicknesses of APLs of gas type, as well as to estimate the value of the reservoir pressure of gas in separate APLs.

14.3.3. On the Possibility of Defining Gas Accumulations in Dense Sandstones

The presence of gas deposits in reservoirs of this type is predicted by geological and geophysical data in the deep horizons in the eastern section of DDB, in the areas of the Belaievsky salt dome and Novomechebilovskaya and Slavyanskaya structures. In the area of the Belaievsky salt-dome structure (Belaievsky-400 well) a relatively large (by area) anomalous zone of the gas accumulation type was detected (Figure 14.5). An area with high values of pressure in the reservoirs was established within its contour. The maximum values of reservoir pressure were estimated by the processing results at $54\,MPa$. This suggests that the gas inflows can be obtained here at depths up to $5400\,m$. The area of the anomalous zone along the contour of 0 is $40\,km^2$, and along the isoline of $50\,MPa$ is $13\,km^2$.

Drilling of the Belaievsky-400 well was completed at a depth of $4446\,m$. The projected drilling depth is $5250\,m$. Eight procedures of fracturing were executed in the well. But commercial inflows of gas were not obtained. In connection with this, additional studies were conducted in the area of the well drilling using the frequency-resonance technology of remote sensing data processing. As a result:

1. The existence of an anomalous zone was validated again. Intense signals (anomalous responses) at the resonant frequencies of free gas were recorded within the mapped anomaly.

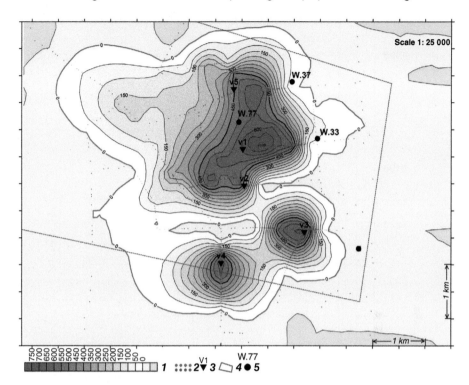

Figure 14.3 Map of anomalous geoelectrical zones of gas deposits type in the area of known gas field. 1 = scale of the FSPEF field intensity (in units of the maximum reservoir pressure, atmosphere); 2 = points of FSPEF survey; 3 = vertical electric-resonance sounding (VERS) points; 4 = prospecting area contours; 5 = wells.

2. Near the Belaievsky-400 well, vertical scanning of remote sensing data was conducted; the following promising intervals for gas detection were identified: (a) 4445–4465 m; (b) 4769–4790 m; (c) 4850–4873 m; (d) 5253–5270 m.
3. Following further processing of remote sensing data, the values of the fluids pressure in the separate reservoir were also assessed: (a) 44.15 MPa; (b) 50.01 MPa; (c) 50.26 MPa; (d) 53.80 MPa.
4. The most promising intervals for gas deposit exploration are numbers 2, 3, and 4 above.

The industrial gas-bearing prospects of the Belaievsky-400 well drilling area may be further assessed using the mobile geoelectrical technology, which includes the ground-based geoelectrical methods of FSPEF and VERS [*Levashov et al.* 2008, 2012c].

However, due to the fact that the Belaievsky-400 well has not been drilled to the projected depth, and, consequently, to the most promising horizons for

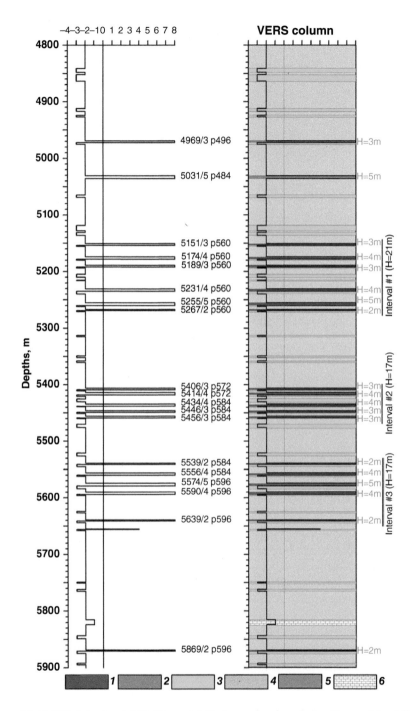

Figure 14.4 VERS data at point V1 (Figure 14.3). Anomalously polarized layer of the following types: 1 = gas; 2 = gas with a small reservoir pressure; 3 = water; 4 = argillite-aleurolite; 5 = sandstone (gas and water collectors); 6 = limestone. 5869/2 p596 = bedding depth of APL of gas type, its thicknesses and value of reservoir pressure in atmospheres, respectively.

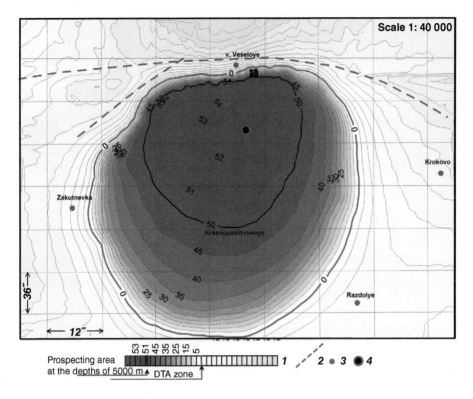

Figure 14.5 Map of anomalous zones of gas deposit type in the area of Belyaevsky-400 well drilling. 1 = scale of the maximum values of reservoir pressure, MPa; 2 = tectonic faults; 3 = settlements; 4 = location of Belaievsky-400 well.

industrial gas accumulation, opening of the projected gas field in Ukraine is postponed indefinitely.

Six anomalous zones of gas and gas-condensate reservoir type of different sizes (in area) and intensity (Figure 14.6) were found and mapped on the site of the Novomechebilovskaya structure's location. Within the largest by area anomalous zone, the maximum value of reservoir pressure is 55 MPa. The area of the anomaly contour along the isoline of 50 MPa is 16 km². Within this zone industrial gas inflows can be obtained at a depth greater than 5000 m. Figure 14.6 shows a second anomalous zone with the maximum reservoir pressure of 55 MPa. However, the area along this anomaly contour of 50 MPa is only 0.8 km². In this regard, when a well location for drilling is provided on the deeper horizons (over 5000 m) in this area, it is advisable to carry out remote sensing data processing over the area on a larger scale.

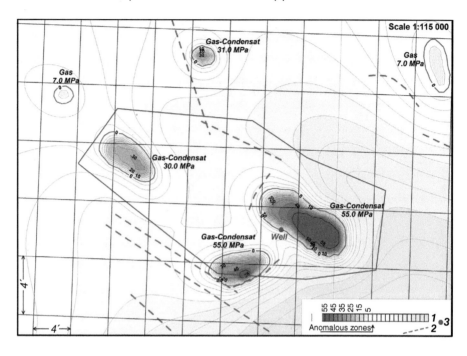

Figure 14.6 Map of the anomalous zones of gas deposit type in the area of Novomechebilovskaya-100 well drilling in Kharkov region of Ukraine (based on the frequency-resonance processing and interpretation of remote sensing data). 1 = scale of the maximum values of reservoir pressure, MPa; 2 = tectonic faults; 3 = approximate position of the drilled Novomechebilovskaya-100 well. Commercial inflows of gas were not obtained in the well.

The total area of all the anomalous zones found within the region of remote sensing data processing is 136.48 km². This is only 5.99% of the surveyed area (2280 km²).

We also note the fact that the construction site of the Novomechebilovskaya-100 well shown on the map (Figure 14.6) (southwest of Alisovka village) falls on the zero isoline of the largest anomalous zone. Consequently, it is not in an optimal location in terms of the results of remote sensing data processing, and hence, the probability of industrial (commercial) gas inflows being obtained after the completion of well drilling is very low.

The Novomechebilovskaya-100 well was drilled to a depth of 3667 m; the commercial gas inflows in the well were not received.

Two anomalous zones of gas and gas-condensate reservoir type with areas of 80 and 8.9 km² (Figure 14.7) have been revealed at the site of the Slavyanskaya structure's location as a result of remote sensing data processing. The area of processed satellite image is 670 km². Regions with relatively high values of

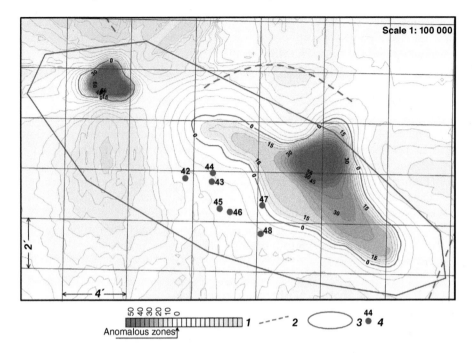

Figure 14.7 Map of the anomalous zones of gas deposit type in the area of Slavyanskaya structure zone (based on the frequency-resonance processing and interpretation of remote sensing data). 1 = scale of the maximum values of reservoir pressure, MPa; 2 = tectonic faults; 3 = approximate contour of the structural zone; 4 = approximate location of the drilled dry wells NN 42, 43, 44, 45, 46, 47, and 48.

reservoir pressure have also been established within the mapped anomalies. Thus, along the 50 MPa contours these areas constitute 7.2 km² (eastern) and 4.5 km² (western).

In Figure 14.7 the blue dots show the location of seven drilled wells within the Slavyanskaya structure, which do not yield a commercial inflow of hydrocarbons. All these wells fall outside the contours of the detected and mapped anomalous zones.

The results of experimental studies allow us to state the following:
1. The frequency-resonance method of remote sensing data processing and interpretation [*Levashov et al.*, 2010, 2011] allows us to detect and map in the areas of unconventional reservoirs—tight sandstones—"sweet spot" zones within which commercial gas inflows may be obtained from drilled boreholes.
2. Application of this method, when prospecting for hydrocarbon accumulations in unconventional reservoirs (carboniferous and crystalline rocks,

shales, tight sandstones), allows us to optimize the location of exploration and production (operating) wells, and thus to significantly reduce their number and their harmful effects on the environment.

3. The drilled Belaievsky-400 well fall on the 53 MPa isoline of the mapped anomaly (Figure 14.5). In this situation, the probability of obtaining commercial gas inflows in the wells is close to 100%.

4. In the areas of the Novomechebilovskaya (Figure 14.6) and Slavyanskaya (Figure 14.7) structures there also are sites within which industrial gas inflows can be obtained from horizons of the cross-section at depths exceeding 5000 m. However, the construction site of the Novomechebilovskaya-100 well (Figure 14.6) falls on the zero isoline contour of the largest anomalous zone. Therefore, the probability of commercial gas inflows being obtained after completion of drilling is close to zero.

5. Seven previously drilled wells within the Slavyanskaya structure (Figure 14.7) do not fall within the contour of anomalies mapped in this area.

14.4. Results of Mobile Method Application in Caspian Basin

14.4.1. License Blocks Atyrau [*Geta et al.*, 2012a, b]

Assessment of the prospects of block (over 10,000 km²) oil-bearing was conducted using the frequency-resonance technology of remote sensing data processing [*Levashov et al.*, 2010, 2011; *Levashov*, 2012]. No anomalous zones with high fluid pressure in the reservoir were found within the Atyrau block as a result of the investigation (Figure 14.8). Consequently, the probability of detecting subsalt deposits there at depths greater than 3000 m is practically zero, and the well with a depth of 7050 m (near the Tasym anomalous zone, southwestern part of the block, Figure 14.8) was drilled in an obviously unpromising location.

The estimates obtained for the maximum values of fluid pressure in the reservoir of 69 MPa in the area of Teren-Uzyuk Eastern structure indicate the possible presence here of subsalt oil deposits (Figure 14.9). Comparison of the mapped anomalies in this area with the anomalous zone over the known Tengiz field (Figure 14.10) indicates the possibility of detection within the Teren-Uzyuk Eastern structure of relatively large hydrocarbon deposits.

In the paper by *Geta et al.* [2012a, Figure 9, p. 104] there is a map of the structures identified in the southern part of the Atyrau block by seismic works. The map also shows the location of several boreholes (including a deep one of 7050 m), from which, according to [*Geta et al.*, 2012a, b], hydrocarbon inflows were not obtained during testing of the wells.

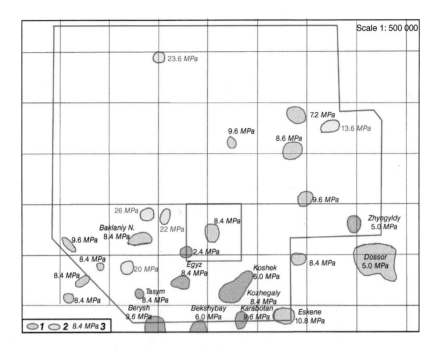

Figure 14.8 Schematic map of the anomalous zones of oil deposit type on the Atyrau license area. 1 = anomalous zones due to APL of oil type in the over-salt column; 2 = supposedly anomalous zones, due to subcornice APL of oil type.

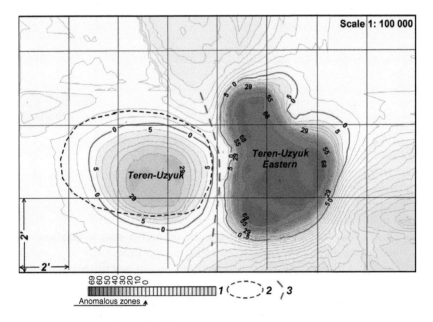

Figure 14.9 Map of anomalous zones of oil deposit type in the area of Teren-Uzyuk oilfield and Teren-Uzyuk eastern structure (based on remote sensing data processing and interpretation). 1 = scale of the maximum values of reservoir pressure, MPa; 2 = approximate contour of the field; 3 = fracture zones.

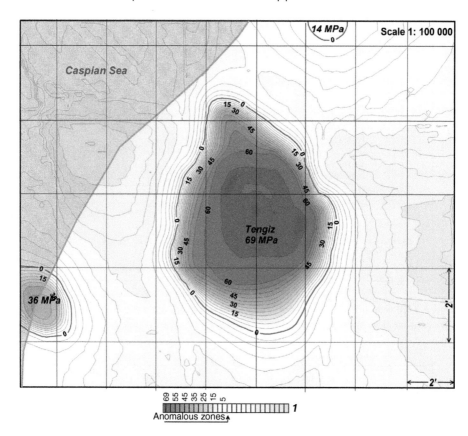

Figure 14.10 Map of anomalous zones of oil deposit type on the Tengiz oilfield (based on the remote sensing data processing and interpretation). 1 = scale of the maximum values of reservoir pressure, MPa.

Comparison of this map with the layout of the detected and mapped anomalous zones of the oil reservoir type showed that the wells drilled here fall outside the contours of the selected anomalies (Figure 14.11).

Joint analysis of the results of remote sensing data processing within the Atyrau block and the findings of similar studies within four major areas in the Northern Turgay region allow us to establish a high probability of detection of industrial hydrocarbon (HC) accumulations in this area and, consequently, the opening of a new oil and gas region in the Republic of Kazakhstan [*Levashov*, 2012]. We can also add that in *Karpov* [2012, p. 3], it is stated that "the prospecting operations conducted in the south of western Siberia (Tyumen region) showed the inconsistency of structural (anticline) rules on wells locations, which indicates the need to change the strategy and methodological approaches to the oil and gas

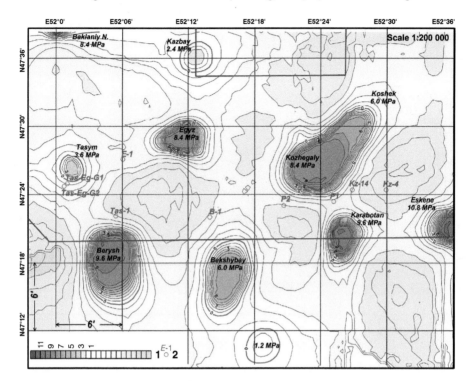

Figure 14.11 Schematic map of anomalous zones of the oil reservoir type with the drilled wells location in the southern part of the Atyrau licence block (Republic of Kazakhstan). 1 = scale of maximum values of fluid pressure in reservoir, MPa; 2 = drilled wells.

prospecting process there (and not only there)". The results of studies in the Caspian Basin are broadly consistent with these conclusions. Thus, within the Atyrau block, 11 boreholes (including one deep) have already been drilled. However, information on HC inflows is lacking [*Geta et al.*, 2012 a, b].

14.4.2. Structure EMBA-B (NUR) Within Block E [*Murzagaliev and Tautfest*, 2012]

The advisability of research on the structure was caused by drilling the NUR-1 deep well (7250 m). In this regard, the results of remote sensing data processing can be certified by drilling (if the well is drilled to the projected depth, of course). So far, it has been drilled to a depth of 5681 m, at a cost of US $39 million. A further US $12–20 million will be needed to drill the NUR-1 well to the projected depth [www.maxpetroleum.com].

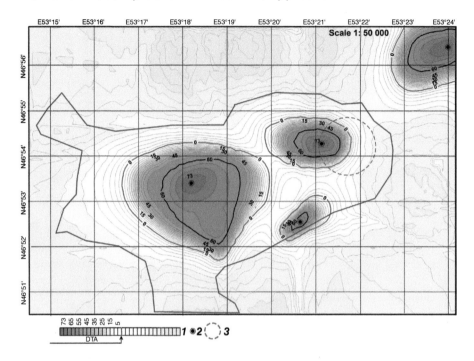

Figure 14.12 Map of anomalous zones of the oil reservoir type within the EMBA-B structure and NUR-1 well of license block E (Republic of Kazakhstan) (based on the remote sensing data processing and decoding). 1 = scale of the maximum values of fluid pressure in reservoir, MPa; 2 = central points of the anomalous zones; 3 = contour within which the well can be positioned.

Remote sensing data of the EMBA-B structure area (Figure 14.12) were processed on a scale of 1:50,000. This is the largest scale of separate area processing compared to the scale of processing within the Atyrau block [*Levashov*, 2012]. The processing results are shown in Figure 14.12. Three oil reservoir type anomalous zones of various sizes were detected and mapped directly within the EMBA-B structure contours. Another anomalous area was fixed outside the EMBA-B structure (Figure 14.12). Assessments of the maximum values of fluid pressure in the reservoir reach 73 MPa within the detected anomalous zones. The decoding results allow us to state the following.

1. The obtained estimations of the maximum values of fluid pressure in the reservoir of 73 MPa indicate a high probability of commercial inflows of oil being obtained from the presalt horizon in the central regions of the detected anomalous zones.

2. The total area of the mapped anomalous zones is substantially less than the area of the EMBA-B structure. The total area of the anomalies fragments

with relatively high values of fluid pressure (over 60 MPa) is even smaller (Figure 14.12).

3.   If the point of the NUR-1 well drilling misses the anomalous area within the contour of 60 MPa, commercial inflows of oil in the well after completion of its drilling to the projected depth will not be obtained.

Within the investigated structure, the following additional studies can be effectively carried out.

1.   The remote sensing data within the area of well location can be processed on a larger scale, 1:10,000 and larger. This will allow refinement of the anomalous contours (and with high fluid pressure in the reservoir also).

2.   In the area of the NUR-1 well, a ground-based survey by the FSPEF and VERS geoelectrical methods may be carried out [*Levashov et al.*, 2008, 2012c]. In addition to clarifying the anomalies contours by FSPEF survey, VERS will also allow us to assess the depth and thickness of APLs of oil reservoir type in the cross-section of the area. The estimates of fluid pressure values in reservoirs in individual APLs of the oil deposit type may also be obtained. Estimates of oil resources in individual APLs of oil reservoir type and throughout the whole cross-section can be calculated when VERS soundings are provided in a sufficiently dense system of points.

3.   The studies listed above in points 1 and 2 can be carried out on all structures within license block E. The materials resulting from such work may be used for optimizing the location of exploratory wells.

4.   The frequency-resonance method of remote sensing data processing and decoding also provides an opportunity to detect and map effectively within the E and A blocks [*Murzagaliev and Tautfest*, 2012] the possible sites of hydrocarbon accumulation in traps of a nonstructural type.

5.   The EMBA-B structure is located close enough (to the north) to the Teren-Uzyuk Eastern structure. Given this, the authors assessed as high the probability of detecting anomalous zones with high fluid pressure in the reservoir in the southern part of license block E. The results shown in Figure 14.12 confirm this hypothesis. Moreover, they show that anomalous zones of the same type can be found in other structures within block E itself, as well as beyond the contours of the structures.

14.5. Pripyat Depression

In late December 2012, Belarus began drilling the 1-Predrechitskaya parametric well, which is within the Rechitsa-Vishanskaya stage of the northern structural and tectonic zone of the Pripyat Trough, with which all Belarusian oil deposits are associated. The Predrechitskaya prospecting area is located between the Rechitsa and Southern Tishkovsky oilfields, thus giving the prospect of assessing the prospects of the search for hydrocarbons as high.

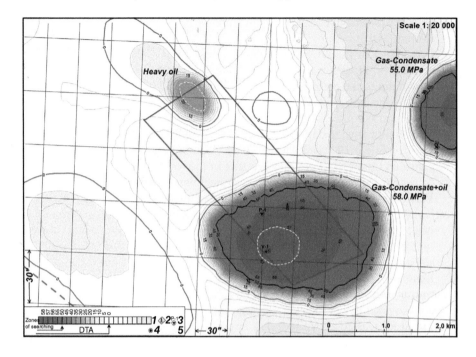

Figure 14.13 Map of the anomalous zones of the hydrocarbon deposit type in a Predrechitskaya prospecting area (Republic of Belarus) (based on the frequency-resonance processing of remote sensing data). 1 = scale of the maximum values of fluid pressure in the reservoirs, MPa (complex parameter); 2 = well; 3 = projected point of the target horizon opening; 4 = "central" point of the anomalous zone; 5 = area of oil detection.

The projected vertical depth of the well is 6481 m, and along the borehole, 6680 m. The planned drilling time was 527 days, and the scheduled time of bringing the facility into operation was the second half of July 2014. The project cost of parametric well drilling was US $27.5 million.

Remote sensing data processing of the area of the 1-Predrechitskaya parametric well was done in November 2013 for a practical demonstration of the suitability of the frequency-resonance technology of remote sensing data processing and interpretation (decoding) [*Levashov et al.*, 2010, 2011, 2012c] for effective assessment of the petroleum potential of both insufficiently studied license blocks and areas, and the deep horizons of the cross-section.

The coordinates of the 1-Predrechitskaya well and the coordinates of two other wells (one of them in production) and the depths of target (productive) horizons in these wells were used as a source of information for processing.

A satellite image of the 1-Predrechitskaya well site location was processed on the scale of 1:20,000. During the processing, the technology of maximum values of fluid pressure in reservoir estimation was used [*Levashov et al.*, 2011].

Within the surveyed area, several anomalous zones of the hydrocarbon deposits type were found and mapped (Figure 14.13).

During the processing of the remote sensing data, the anomalous responses on the resonant frequencies of gas, gas condensate, and oil were recorded. At the same time, we note that the resonant frequencies of light oil and gas condensate are very close. The "area of oil detection" in Figure 14.13 is an area where more clearly and confidently anomalous responses on the resonant frequencies of oil were fixed directly. In the northern part of the survey area, an anomalous zone of heavy oil type was recorded. This is due to the fact that within its contours the anomalous responses at resonant frequencies of paraffin were fixed.

The 1-Predrechitskaya well falls into the anomalous zone within the contour of 55 MPa (Figure 14.13). Therefore, we can assume that it is located in a relatively optimal location (based on remote sensing data processing, of course), and the likelihood of industrial (commercial) HC inflows being obtained after completion of its drilling is very high. Nevertheless, the optimal (central) point of the anomalous zone is located to the south (Figure 14.13).

Note also that the remote sensing data processing of a larger fragment of the territory was done on a scale of 1:30,000. As a result, four anomalous zones of the hydrocarbon deposits type that were similar in size and intensity were found and mapped within the surveyed area. A fragment of one of them is shown in the eastern part of Figure 14.13.

In the area of the 1-Predrechitskaya well, ground-based geoelectrical research with the FSPEF and VERS methods may also be carried out [*Levashov et al.*, 2008, 2012c]. In addition to clarifying the contours of the anomalous zones by FSPEF survey, VERS will also assess the depth and thicknesses of APLs of the oil deposit type in the cross-section of the area. From this, the values of reservoir pressure in the individual APLs of the oil deposit type can be obtained.

It is worth noting that the results of ground geoelectrical research are superior in accuracy and detail to the results of remote sensing data processing. By carrying out VERS over a dense system of points, we can also obtain estimates of oil resources in the individual APLs of oil deposit type, and throughout the cross-section as a whole.

14.6. Conclusions

The results of the application of the mobile technology in the different oil- and gas-bearing region, presented above as well as in *Levashov* [2012], indicate the possibility of using them effectively to recommend hydrocarbon exploration based on new and, most importantly, independent information, to help accelerate and optimize the exploration process as a whole.

Comparison of the mapped anomalous zones with the approximate contours of the structures investigated in the different region demonstrates the value and importance of the new information obtained by the frequency-resonance method of remote sensing data processing and decoding. This additional information allows us to significantly reduce (localize) the areas of priority search for hydro-carbon accumulations. It is advisable to use the innovative method of frequency-resonance processing and decoding of remote sensing data more actively within the traditional complex of oil and gas exploration.

The above information indicates the feasibility of conducting a wide range of geological and geophysical exploratory work, when choosing the locations of exploration wells. Given the results of testing the technology of frequency-resonance processing and decoding of remote sensing data, it is useful to note the following.

1. Within the block studied by the 2-D and 3-D seismic surveys, the detected and mapped structures may be promising areas that are not in the structures' anticline and have therefore not been drilled.

2. Such areas can be detected and mapped effectively using the frequency-resonance method of remote sensing data processing.

3. Remote sensing data processing of the seismically studied blocks will provide more complete information about their prospects for oil and gas. As a result of such work, the areas of potential hydrocarbon accumulations in traps of a nonstructural type can also be effectively identified.

To determine the optimal locations for exploration wells (especially deep ones, including for subsalt deposits), the ground-based geoelectrical methods FSPEF and VERS may be used [*Levashov et al.*, 2008, 2012c]. Recently, these have been substantially improved. Thus, the FSPEF method allows us also to select and map anomalous zones with higher values of fluid pressure in the reservoir. Most importantly, the improved VERS method allows assessment of the value of fluid pressure in the reservoir in separate anomalous polarized layers of oil and gas types.

The authors have repeatedly pointed out [*Levashov et al.* 2012c, d] that the nonclassical FSPEF and VERS methods have contributed significantly to the development of a new "substantial" ("material") paradigm of geophysical research, within which the direct search is conducted for a specific physical sub-stance (material): gas, oil, gas hydrates, water, ore minerals, and rocks (gold, platinum, silver, zinc, uranium, diamonds, kimberlites, etc.). The first stage of development of this paradigm can be considered the initial research and development of direct methods for oil and gas exploration. At the same time, in geological-geophysical terminology, the well-known and currently widely used (including by the authors) expression, the "deposit" type anomaly, was intro-duced. There is evidence that the effectiveness of geophysical methods based on the principles of the substantial paradigm is greater than that of traditional methods.

We also note that a significant contribution to the establishment of the substantial (material) paradigm of geophysical research is also made by the frequency-resonance technology of remote sensing data processing and interpretation, the practical approbation of which has been carried out by the authors since 2010. This technology is focused on the discovery and mapping by satellite data of anomalies of the oil reservoir, gas accumulation, aquifer, zone of gold mineralization types, etc. The combined use of remote sensing data processing and interpretation and the FSPEF-VERS technologies at different stages of geological and geophysical studies allows us to significantly optimize and speed up the search and exploration stages of geophysical surveys.

Finally, we note again that the results of each new practical application of the frequency-resonance technology of remote sensing data processing reinforce the value of the following concluding statements, which have been repeatedly cited by the authors in various documents and publications [*Levashov*, 2012].

Service companies in the geosciences field may be interested in applying the mobile frequency-resonance technology of remote sensing data processing in the initial (reconnaissance) stages of prospecting, when conducting exploratory geological and geophysical work on specific prospective areas and sites. Using this technology will allow them to receive operatively additional (and, more importantly, independent) information on the petroleum potential of the studied areas. Anomalous zones identified and mapped in the future can be more thoroughly studied by traditional (seismic, primarily) geophysical methods. In general, this will allow us to more soundly and confidently identify prospective sites for the location of exploratory wells.

Oil companies, operators of specific license areas and blocks, should also be interested in using the frequency-resonance technology of remote sensing data processing at different stages of prospecting and exploration. Its use at the initial stages of prospecting will allow 3-D seismic surveys of increased detail to be conducted within detected and mapped anomalies of the HC deposits type. Application of the technology for further evaluation of the petroleum potential within identified seismic structures and objects will optimize the layout of the first exploratory wells. In general, application of the technology will allow the exploration process to be accelerated and optimized.

The results of remote sensing data interpretation provide new and mainly independent information that can be useful and marketable at any stage of the oil and gas exploration process. An important characteristic feature of this information is the efficiency with which it can be acquired. Applying the method of assessing the maximum value of fluid pressure in a reservoir allows us to more substantially narrow down the areas of priority for detailed field prospecting and the location of exploration wells. Mobile geophysical technologies allow us to rapidly identify and map potential accumulations of oil and gas in both traditional and nontraditional types of reservoir.

Effective remote sensing data processing to determine areas for HC prospecting and well drilling provides a significant amount of additional and independent information that, together with available geological and geophysical data, allows us to create a more complete idea of the prospects of their oil and gas content. The independent nature of this information lies in the fact that it was obtained without the involvement of the available geological and geophysical materials from studies of previous years. And, most significantly, the financial and time expenditure required to obtain this information is far less than the costs (time and financial) of geological and geophysical study within the license areas in the past.

Another special feature of the received data is that it can be classified as direct signs of oil and gas bearing at the surveyed sites. The resulting materials indicate more clearly the presence of HC accumulations in the cross-section. They delineate more specifically areas for HC deposits prospecting and significantly narrow the field for optimally locating exploratory wells.

Using these mobile methods and technologies, which allow us to obtain new information with direct signs of oil and gas potential of hydrocarbon accumulations in unconventional reservoirs, may reduce significantly the number of prospecting, exploration, and production wells and, hence, their negative impact on the environment. In Ukraine (as in other European countries) environmental issues during the development of hydrocarbons in unconventional reservoirs are crucial.

Acknowledgments

The authors are deeply grateful to three anonymous reviewers whose comments and suggestions helped to improve the manuscript's structure and contents.

References

Geta, S., et al. (2012a), Exploration of the northern slope of Aktobe-Astrakhan uplifts system on the Atyrau block, Caspian depression, actual problems of geology and petroleum potential. First International Geological Conference AtyrauGeo-2011, Atyrau, Republic of Kazakhstan, September 2011. *Proceedings of the ONGC, 1*, Atyrau 2012, 96–106 (in Russian).

Geta, S., et al. (2012b), Oil and gas prospects of over-salt Triassic sediments on the Atyrau block, Caspian depression, actual problems of geology and petroleum potential. On the basis of the First International Geological Conference Atyrau Geo-2011, Atyrau, Republic of Kazakhstan, September 2011. *Proceedings of the ONGC*, 1, Atyrau 2012, 183–187 (in Russian).

Karpov, V. (2012), Status and prospects of oil and gas exploration in Western Siberia. *Oil and Gas Geology*, 3, 2–6 (in Russian).

Kovalev, N., et al. (2009), Use of geoholographic complex "Poisk" for detecting hydrocarbons. *Geoinformatics*, 3, 83–87 (in Russian).

Levashov, S. (2012), Application of mobile geophysical technologies for petroleum potential evaluation of large blocks and deep horizons of cross-section (Caspian Basin, Republic of Kazakhstan). *Geoinformatics*, 4, 5–18 (in Russian).

Levashov, S., et al. (2008), Express-technology of direct prospecting and exploration of hydrocarbon accumulations by geoelectrical methods. *Oil Industry*, 2, 28–33 (in Russian).

Levashov, S., et al. (2010), New possibilities of the operative estimation of oil-and-gas prospects of exploratory areas, difficult of access and remote territories, license blocks. *Geoinformatics*, 3, 22–43 (in Russian).

Levashov, S., et al. (2011), Assessment of relative value of the reservoir pressure of fluids: results of the experiments and prospects of practical applications. *Geoinformatics*, 2, 19–35 (in Russian).

Levashov, S., et al. (2012a), Experience of geoelectrical methods using the hydrocarbon potential study of deep horizons in DDB. *First International Hydrocarbon Conference: Potential Great Depths, Energy Resources of the Future—The Reality and Prognosis, Baku-2012*, Nafta Press, 79–82 (in Russian).

Levashov, S., et al. (2012b), On the possibility of using the values of fluid pressure in the reservoir for evaluation of the hydrocarbon potential of deep horizons in cross-section. *First International Conference Hydrocarbon: Potential great depths, energy resources of the future—The reality and prognosis, Baku-2012*, Nafta Press, 69–72 (in Russian).

Levashov, S., et al. (2012c), Frequency-resonance principle, mobile geoelectrical technology: a new paradigm of geophysical investigation. *Geophysical Journal*, 34(4), 167–176 (in Russian).

Levashov S., et al. (2014), Mobile geophysical technologies: Experimental study of possibility of application for hydrocarbon accumulations prospecting within areas of shale spreading in Eastern Europe. *Geoinformatics (Ukraine)*, 4, 5–29 (in Russian).

Levashov S., et al. (2015), Mobile direct-prospecting methods: New opportunities of oil and gas exploration accelerating and optimization. *Oil and Gas*, 2, 93–115 (in Russian).

Murzagaliev, R., and J. Tautfest (2012), A promising new area of petroleum accumulation in Paleozoic sediments on the southern slope of Gurievsky arch, Caspian depression: Actual problems of geology and petroleum potential. On the basis of the First International Geological Conference AtyrauGeo-2011, Atyrau, Republic of Kazakhstan, September 2011. *Proceedings of the ONGC*, 1, Atyrau 2012, 90–95 (in Russian).

Pukhliy V., et al. (2010), *Nuclear Magnetic Resonance: Theory and Applications*. Cherkasky CSTEI, Sevastopol (in Russian).

Rostovtsev, V., et al. (2011), Great oil deposits of Russia. *Geomatics*, 1, 60–63 (in Russian). http://www.geomatica.ru/eng/.

Shuman, V. N. (2012), Electromagnetic-acoustic conversion and high-sounding system, new opportunities and new formulations of old questions. *Geophysical Journal*, 34(3), 32–39. (in Russian).

Solovyov, V., et al. (2011), Gas hydrates accumulations on the South Shetland Continental Margin: New detection possibilities. *Journal of Geological Research*, Article ID 514082. doi:10.1155/2011/514082.

Weaver, B. W., and R. K. Warren (2004), Electric power grid induced geophysical prospecting method and apparatus. International Patent No WO 2004/106973 A2, Dec. 9, 2004.

Yakymchuk, N., et al. (2008), Express-technology for direct searching and prospecting of hydrocarbon accumulation by geoelectrical methods. *International Petroleum Technology Conference*. Kuala Lumpur, Malaysia. Paper IPTC-12116-PP. https://www.onepetro.org/conference-paper/IPTC-12116-MS.

Yakymchuk, N. A. (2014), Electric field and its role in life on Earth. *Geoinformatics (Ukraine)*, 3, 10–20 (in Ukrainian).

Yakymchuk, N., et al. (2015). Mobile technology of frequency-resonance processing and interpretation of remote sensing data: The results of application in different region of Barents Sea. *Offshore Technology Conference*. doi:10.4043/25578-MS. https://www.onepetro.org/conference-paper/OTC-25578-MS.

15. ANOMALIES OF LOW DENSITY IN THE CRYSTALLINE CRUST OF THERMOBARIC ORIGIN: A NEW INSIGHT INTO MIGRATION AND LOCALIZATION OF HYDROCARBONS

Valery Korchin

Abstract

This chapter presents a new hypothesis of the formation, existence, and disappearance of low-velocity zones (LVZs) discovered by the deep seismic sounding (DSS) method as velocity anomalies of the Earth's crust. The increase of geological mapping depths and the study of the possible localization of mineral deposits require a new level of understanding of the geological structure of the studied areas, as well as their tectonic-magmatic and structural development, to substantially extend the prospects of applying geophysical methods and, overall, to promote the efficiency of prospecting and surveying works. This is closely related to the study of the physical properties of mineral material at the different PT regimes. The different changes of $V_{P,S}=f(PT)$, $\rho=f(PT)=f(H)$ for rocks are experimentally set up (regions of velocity inversion and density are revealed). There is a cataclastic increase in volume owing to rearrangement and recrystallization of constituent of the mineral medium: the microdilatancy phenomenon among them. The configuration and location of LVZs correlate well with DSS elastic anomalies. Zones arise when the temperature gradient exceeds a certain threshold and the pressure cannot compensate for the disturbance of rocks caused by temperature effects. The presence of LVZs is an objective reality, and they can be found everywhere in the Earth's crust. LVZs' thermal and baric instabilities are stipulated by their fragmentary occurrence in the Earth's crust due to their vertical and horizontal migration, depending on temperature fluctuation.

The physical characteristics of different types of crystalline rocks largely depend on their fractures and porosity, as well as the state of their

Institute of Geophysics of Ukraine National Academy of Science, Kiev, Ukraine

Oil and Gas Exploration: Methods and Application, Monograph Number 72,
First Edition. Edited by Said Gaci and Olga Hachay.
© 2017 American Geophysical Union. Published 2017 by John Wiley & Sons, Inc.

intergrain boundaries. Interest in research into rock imperfections has increased considerably due to studies of the migration of gas-liquid fractions, in particular hydrocarbons, in crystalline rocks at different depths of the lithosphere. It is practically impossible to expose the whole gamut of appropriate and causal reasons that influence the formation of pores and microfractures. However, a tendency has been found that in petrographically similar groups of rocks, the increase of porosity and microfractures leads to a reduction of the velocity of elastic waves in them.

Based on the present laboratory study of the relationship between density and the longitudinal velocity of mineral material at high pressures and temperatures, zone anomalous behavior of porosity and fracture parameters exist at different depths in crystalline crust, which can be canals for the migration and localization of hydrocarbons of deep origin.

15.1. Introduction

Today, there are 469 oil and gas commercial fields in crystalline basements of 29 countries [*Kutcherov and Krayushkin*, 2010]. Among them are 55 gigantic and super gigantic fields, which contain 18% of the total world proven reserves of oil and 5.4% of the total world proven reserves of natural gas. As low porosity and permeability are characteristic of crystalline rocks in basements, the quality of hydrocarbon reservoirs depends on the value of their secondary porosity, which is usually explained by shrinkage during cooling, occurring joints, fractures and faults, and weathering and dissolutioning effects of hydrothermal circulation [e.g., *Sircar*, 2004; *Han et al.*, 2013; *Anderson et al.*, 2015].

The aim of this chapter is to present a new mechanism for developing porosity in the crystalline crust and extend kinds of traps for deep abiogenic hydrocarbons. The best candidate for this is LVZs due to the similarity of their velocities to those of fractured rocks in hydrocarbon fields. The LVZs can be oil and gas reservoirs if only hydrocarbons are transported to them from deep sources along pathway systems of different types [*Kun, Suyun*, 2015]. The origin of deep abiogenic hydrocarbons and their migration into the Earth's crust are comprehensively reviewed in [*Kutcherov, Krayushkin*, 2010].

Deep seismic sounding (DSS) has registered LVZs at a depth of 3–15 km in the Earth's crust of the Precambrian shields all over the world (Figure 15.1). Long-term experimental studies of samples with different mineral composition under high pressure and temperature have revealed complex relations between elastic parameters and thermobaric factors [*Korchin*, 2012; *Korchin*, 2013a, b; *Korchin, Burtny and Kobolev*, 2013]. The essence of these experiments is to determine the velocity-density characteristics of samples by the programmatic influence of *PT* values corresponding to those at different depths in the lithosphere.

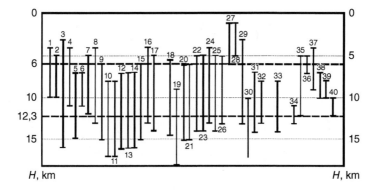

Figure 15.1 The waveguides in the crust of shields: the Ukrainian (1–17), Baltic (18–32), Indian (33), and Canadian (34–40) [*Tripolsky and Sharov*, 2004]. ΔV_p = 0.1–0.7 km/s (0.1–0.22 km/s); (diminishing of velocity in zone), and ΔH = 3–15 km; H_{min} = 5–12 km (depth of bedding of minimum V_p in the low-velocity zone).

The velocity inversion is clearly seen on the experimental curves of $V_{P.S}=f(PT)=f(H)$. Some increase of $V_{P.S}$ and ρ with increasing depth (thermobaric parameters) is followed by a decrease in these parameters. Then the velocity and density increase again (Figure 15.2).

Thus, an LVZ is formed on the $V_{P.S}=f(PT)=f(H)$ curve whose configuration and location correlates well with DSS information from the Earth's crust (Figure 15.1). Changes of $V_{P.S}=f(H)$ can also be obtained from measuring data of isobars of velocities ($V_p=f(T)$) at $P=const$ and their isotherms ($V_p=f(P)$) at $T=const$) (Figure 15.2b). Detailed studies have shown that both methods of determining $V_p=f(H)$ lead to identical results [*Korchin*, 2013a].

15.2. Influence of the *PT* Regimes on Elastic Characteristics of Rocks

Regardless of the methods of determining velocities, using several *PT* programs or by means of calculating isobars and isotherms, there is a threshold value of changes of temperature with depth $\left(\dfrac{\partial T}{\partial H}\right)$ when the anomalous elastic state of mineral material arises (zones of low velocities). The changes of velocity of elastic waves with depth (V_p) in a rock of constant mineral composition can be calculated using the ratio $\dfrac{\partial V}{\partial H}=\left(\dfrac{\partial V}{\partial P}\right)_T\dfrac{\partial P}{\partial H}+\left(\dfrac{\partial V}{\partial T}\right)_P\cdot\dfrac{\partial T}{\partial H}$. Zones of low velocity in the Earth's crust are determined by the assumption of $\dfrac{\partial V}{\partial H}<0$. As far as $\left(\dfrac{\partial V}{\partial P}\right)_T,\dfrac{\partial P}{\partial H},\dfrac{\partial T}{\partial H}$ are positive, and $\left(\dfrac{\partial V}{\partial T}\right)_P<0$, then to form a zone it is necessary to fulfill the assumption for absolute values:

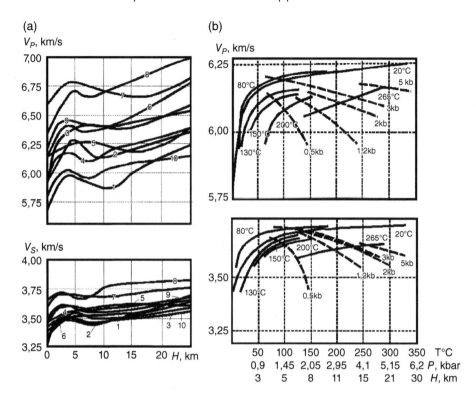

Figure 15.2 (a) Changes of $V_{P,S} = f(PT)$ with depth: 1 = even-grained granites, 2 = porphyry granites, 3 = rapakivi granites, 4 = plagiogranites, 5 = trachytoid granites, 6 = middle rocks, 7 = basic rocks, 8 = ultra basic rocks, 9 = charnokitoides, 10 = gneisses. (b) Isobars and isotherms of change velocity of V_p and V_s in granite.

$$\left| \left(\frac{\partial V}{\partial P} \right)_T \cdot \frac{\partial P}{\partial H} \right| < \left| \left(\frac{\partial V}{\partial T} \right)_P \cdot \frac{\partial T}{\partial H} \right| \qquad (15.1)$$

In most cases in the Earth's crust, the change of lithostatic pressure with depth can be considered a constant value. Therefore, $\left(\dfrac{\partial P}{\partial H} \right)$ is 0.24–0.32 kbar/km at depths from 3 to 40 km for ancient shields. The gradient of temperature changes at these depths varies within wide limits (from 5 to 25 °C/km) [Korchin et al., 2013; Kutas, 1978]. Studies have shown that the relative increase of velocities with pressure for room temperatures is characterized by two areas: $P = (0–2)$ kbar is the area of maximum increase of velocity, $P > 2$ kbar where the minimum gradient of velocity changes occurs. As a rule, the velocity change with temperature for atmospheric pressure has three areas: $T < 80–100$ °C (minimum changes), $T \approx 80–250$ °C (maximum change of V_p).

Further heating within the interval of $T = 250–600\,°C$ leads to a low decrease of the velocity. Relative changes of velocity under compensating constant pressure (isobars) and constant external temperature (isotherms) differ by absolute values. Within the interval of $20–70°4C$ under $P < 0.7\,kbar$, the changes of velocity with temperature are negligible, i.e., up to the depth of $2–3\,km$ the velocities always increase intensively. This is caused by the increase of V_p with pressure due to in duration of rocks. The interval $T = 100–250\,°C$ is the area with the most intensive changes of $V_p = f(T)$. Here, a twofold decrease of velocity is possible because of the influence of temperature under atmospheric pressure, and a decrease by about 10%–20% under compensating pressure of $P \approx 1–4\,kbar$. It is in this interval of pressures and temperatures ($P \approx 1.2–3.5\,kbar$; $T \approx 110–250\,°C$) that the greatest negative changes of the velocity of elastic waves propagation are observed and zones of low velocities are revealed. Experimental studies have found $\left(\dfrac{\partial V}{\partial T}\right)_P = -2.7 \pm 0.5$ ($P \approx 0.5\,kbar$); -0.7 ± 0.3 ($P = 2\,kbar$); $-0.33 \pm 0.1\,m/s \cdot °C$ ($P = 5\,kbar$) and V_p from pressure at different constant temperatures $\left(\dfrac{\partial V}{\partial P}\right)_T = 0.8 \pm 0.3$ (pressure interval 0–2 kbar, temperature 20–80 °C); 0.01 ± 0.005 (under $P \approx 2–5\,kbar$, $T \approx 20–80\,°C$); 0.04 ± 0.01 (at $P \approx 2–5\,kbar$, $T \approx 265\,°C$) [*Korchin et al.*, 2013]. Based on these data and experiments (low- and high-temperature regimes), calculations have shown from the realization of low-temperature regime experiments that $\left(\dfrac{\partial T}{\partial H} < 9 - 11\dfrac{°C}{km}\right)$ zones of velocity inversion are not reflected in the relationship curves of $V_p = f(PT)/f(H)$. If the temperature gradient is $\left(\dfrac{\partial T}{\partial H} > 15 - 20\dfrac{°C}{km}\right)$ in the pressure interval 1.8–3.5 kbar, the relationships of $V_p = f(H)$ clearly reveal low-velocity zones (Figure 15.3a). The decrease of velocities in these zones of different samples varies from –10 to –250 m/s (Figure 15.3b).

The thickness of the LVZs is from 2 to 20 km (60% in the interval of 6–12 km). As a rule, the depths of the minimum values of V_p are located in the interval of superposition of pressures and temperatures, corresponding to 6–21 km (Figures 15.2 and 15.3). Table 15.1 presents the mean values of parameters characterizing the LVZs for the rocks studied.

As is seen (Figures 15.2, 15.3; Table 15.1), the configurations of the experimental LVZs (depth, thickness, velocity decrease) are similar to those discovered from the DSS results. On the curves of $V_p = f(PT) = f(H)$, three areas are observed (Figure 15.3, Table 15.1). The first area corresponds to the initial interval of depths (0–5 km), where the velocities of V_p and V_s increase sharply. Here intensive compression of rocks occurs with depth due to the lithostatic pressure: the most opened microscopic cracks and some pores are closed in the rock by compression influence. Special attention has been focused on the study of the relationship between the gradients of the elastic velocities and the mineral composition in the inversion zones of $V_p = f(PT) = f(H)$.

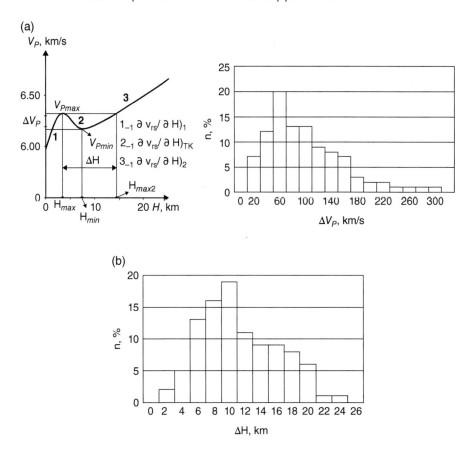

Figure 15.3 (a) Basic parameters of low-velocity zones (Table 15.1). (b) Distribution histograms of ΔV_p (diminishing of velocity in zone) and ΔH (depth of bedding of minimum V_p in the low-velocity zone).

15.3. Influence of Mineral Composition of Rocks on Their Elastic Characteristics

Rocks with increased content of quartz result in ultravelocity zones in comparison with other mineral associations. The highest values of gradients of lowering velocity in LVZ are observed in porphyry granites and rapakivi granites as well as in biotite gneisses, which have low initial values of elastic velocities. As a rule, the higher the initial velocities in the rocks, the higher the ΔV_p and ΔH. This is evidence that the less the rock structure is destroyed, the more it is subjected to transformations in the conditions of programming high pressures and temperatures in the inversion zone in laboratory experiments.

Table 15.1 Mean values of parameters characterizing the area of low-velocity zones (LVZs).

Parameters Rocks	ΔV_p, km/s	ΔH, km/s	V_{Po}, km/s	V_{Pmax}, km/s	H_{max}, km	V_{Pmin}, km/s	H_{min}, km	$\left(\dfrac{\partial V}{\partial H}\right)_{LV}$	$\left(\dfrac{\partial V}{\partial H}\right)_2$
Even-grained granites	0.11	9	5.6	6.08	4.46	5.97	10.5	−0.022	0.023
Rapakivi granites	0.11	10.6	5.83	6.36	7.05	6.25	13.7	−0.029	0.055
Porphyry granites	0.15	12.1	5.91	6.26	6.21	6.11	13.21	−0.022	0.030
Plagiogranites	0.08	10.5	5.92	6.16	4.79	6.13	6.36	−0.018	0.012
Trachytoid granites	0.08	13.8	6.07	6.52	5.67	6.41	15.77	−0.010	0.013
Diorites	0.07	8.5	6.16	6.41	5.75	6.34	9.75	−0.016	0.015
Labradorites	0.1	11.1	6.53	6.69	4.63	6.59	10.75	−0.021	0.022
Gabbros	0.13	11.7	6.76	6.76	5.58	6.63	11.92	−0.018	0.022
Pyroxenite	0.09	9.5	6.33	6.41	5.75	6.34	9.75	−0.018	0.020
Charnokitoides	0.11	10.2	6.27	6.56	3.86	6.45	9.33	−0.014	0.028
Biotite gneisses	0.08	9.4	5.51	5.68	4.39	5.61	8.88	−0.027	0.033
Pyr-Amph gneisses	0.09	10.3	6.09	6.15	7.84	6.06	13.81	−0.018	0.019

Note: ΔV_p = decreasing velocity in LVZ; ΔH = zones power; V_{Po} = velocity at the atmospheric terms; V_{Pmax}, H_{max} = first maximum and its depth; V_{Pmin}, H_{min} = minimum velocity in a zone and its depth; $(\partial V_p/\partial H)_{LV}$ = gradient of velocity changes in LVZ; $(\partial V_p/\partial H)_2$ = intensity of growth of velocity after LVZ.

High pair coefficients of correlation are observed between the minimum values of velocities of elastic waves in the inversion zone (V_{Pmin}) with the content of the main rock-forming minerals in all groups of rocks studied. Such a correlation is determined between the minimum values of LVZ depth (H_{min}) and the quantity of quartz. It is possible to suppose that in microdeforming processes and the lowering of mineral density in low-velocity zones, grains of quartz are the most active. These grains seem to result in brittle shifting dislocations of individual domains of contiguous minerals in the complex intense state of a rock sample under certain *PT* conditions [Korchin et al., 2013; Korchin, 2006].

An analogical result has been obtained by the analysis of multiple cross-correlations between ΔV_p and ΔH relationships and the percentage of the main rock-forming minerals. The basic minerals influencing on parameters for granitoids and gneisses are quartz and plagioclase content. An increase of the content of quartz and plagioclase in acid rocks and gneisses leads to the decrease of ΔV_p and ΔH values. However, the increase of potassium feldspar may decrease or increase these parameters.

The velocity of changes in the range of $H > 18\,km$ is mostly influenced by the plagioclase content and much less by the biotite and pyroxene percentage. Most groups clearly demonstrate that the lower the velocity of elastic waves in LVZs, the steeper their gradients with the increase of depth. In all cases, the gradient change of the V_P velocity is much steeper than for the V_S velocity.

An analysis of multiple correlation coefficients (R) between all the different possible combinations of main rock-forming minerals and velocities shows that a tendency for the influence of main rock-forming minerals on elastic parameters to decrease with depth is characteristic of groups of rocks. There are increasing multiple correlation coefficients in the first area (up to 3 km depth). In the inversion zones (3–16 km depth), the multiple correlation coefficients decrease, while outside the LVZ they increase again, although they remain less than in the first area of small depths.

Such regularities demonstrate that the influence of PT parameters from depth (within the Earth's crust) does not result in increased differentiation of the mineral composition of the rocks and elastic parameters. In the low-velocity zones, the influence of mineral composition on the elastic parameters of rocks is negligible. It is considerably less than the structural transformations of rocks under pressure and temperature.

15.4. Nature of Low Velocity Zones: Thermobaric Decrease of Rock Density

Based on the results of integrated structural studies of different rocks from the Ukrainian Shield at high pressure and temperature by optical, X-ray, and microprobe methods, and the analysis of determinations of elastic characteristics, the elastic vertical zoning of mineral material has been proposed in the Earth's crust (Table 15.2) [*Korchin*, 2013a; *Korchin et al.*, 2006; *Korchin*, 2006].

Elastic and structural changes of rocks under thermodynamic conditions at depths of 3–5 km to 12–15 km are characteristic of the properties of rocks during their cataclastic transformations. Within this interval of density lowering of the mineral medium, there occurs a dilatancy phenomenon. The main mechanism for this phenomenon is the common action of heterogeneous strains within the volume of the sample, which in local contacts sometimes reach values exceeding the limits of strength of individual minerals, which leads in turn to brittle micro-destruction of matter. Such density lowering results from differently oriented anisotropic coefficients of linear expansion of individual minerals, the effect of destruction of gas-liquid inclusions, and the migration of free water and gas along the microfissures of rock. Microstructural disturbances of rock prepare the conditions for its essential transformation with the increase of depth (PT conditions of experiment). Gradually, mechanisms for plastic deformations of the medium arise, and improvement of the rock occurs by means of material and structural (at the level of elementary defects) reconstructions. In geological terms, within the interval of depths of 20–40 km, the process of regional metamorphism appears to occur.

Table 15.2 Microstructure transformations at different pressures and temperatures.

Microstructure Characteristics and Properties	Parameters of PT Influence and Determined Changes	
	LVZ 2–3 kbar,160–220°C, H = 3–15 km	Area of Velocity Increase 5.5–6.5 kbar, 300–350°C, H > 10–15 km
Vp, Vs, E, G, σ, K, ρ, $1/\beta$	Decreased	Increased
Twinning and mylonitization	Considerably	Small, diminished
Expansion of intergrain boundaries	Considerably increased	Decreased, boundaries made more compact
Main microcracks	Increase	Diminished, absent
Change of the gas-liquid inclusions	Splitting, destruction	Decreasing in volume, localization
Optical anisotropy	Increased	Decreased
Polarized optical heterogeneity	Increased	Decreased
Curvature of biotite plates, double bars of plagioclase	No	Exposed
Reorientation of grains of quartz, biotite, plagioclase	No	Exposed
Lines of biotite gliding	Exposed	Decreasing of lines
Change of form of quartz grains	Splitting	Turning, recrystallization
Mosaic blocks	Decreased	Increased
Density of dislocations in blocks	Increased	Decreased
Density of dislocations in intergrain boundaries	Decreased	Increased
Relative deformation of grains	Increased	Decreased, change of sign
Packing defects of twins	Increased	Decreased
Density of dislocations of individual grains	Increased	Decreased
Character of the dislocation field, dynamics	Increase of centers of generation, translation sliding	Creeping, annihilation, localization, increase of dislocation bars, conservative sliding
Character of deformation	Low-temperature elastic work-hardening, fragile destruction	Elastic and plastic high-temperature compression

Thus, low velocity zones in the Earth's crust are considered to be an objective reality that results from structural transformations of rocks as a consequence of the action of pressure and temperature at relevant depths in the lithosphere. These arise when the temperature gradient at relevant depths exceeds a certain threshold and the pressure is unable to compensate the disturbance of the mineral medium produced by temperature.

15.5. Influence of Pressure and Temperature on Rock Density

Based on comprehensive experiments with rocks and minerals under different PT conditions simulating those in the LVZ, the relative deformation of grains and their twinning increase in the LVZ (5–15 km depth). The density increases inside the blocks' dislocations and decreases on the intergrain boundaries. Defects in the mineral packing increase. Intergrain boundaries expand due to their mylonitization, and the amount of main microfissures increases. There is depressurizing or opening of the gas-liquid inclusions in the minerals under heating because of the excess pressure. Moreover, the rocks are characterized by decreased density (Figure 15.4).

Information on density versus depth changes has been obtained from studies of a decrement in volume under PT experiments and ultrasonic determinations of the compressibility of rocks. The density characteristics of rocks change nonlinearly with depth, similar to the elastic parameters.

The $\rho = f(PT) = f(H)$ relationships demonstrate maxima and minima. This implies that the deep simultaneous influence of P and T on the mineral material results in the formation of the LVZs. This inference of the mechanism for the origin of the LVZ is supported by studies of rock density programming the influence of P and T. As expected, density also decreases under PT relative to the LVZ.

The values of $\partial \rho / \partial H$ are sometimes negative in conformity with the considerable density lowering of rocks that creates the zones of low density in the crust. Under thermodynamic conditions corresponding to 5–15 km depth, the gradient of density increment falls to zero or becomes negative (Figure 15.4). Additional experiments have revealed that LVZs depend weakly on the mineral composition of the crust. Obviously, structural transformations and chemical composition play a key role in the variations in rock density.

The horizons of the low density are very sensitive to the temperature conditions of the Earth's crust, similar to the LVZ. Increasing deep heat flow reduces the rock density, activates the capability to lower the density, and increases the permeability and hygroscopicity of rocks (activating a process fluid movement), which leads to the metamorphic transformation of rocks. In other words, these LVZs are the most active horizons of modern geological-geophysical transformations of the mineral environment of the crystalline crust [*Korchin*, 2012; *Korchin*, 2013a, b; *Korchin et al.*, 2006; *Korchin*, 2006].

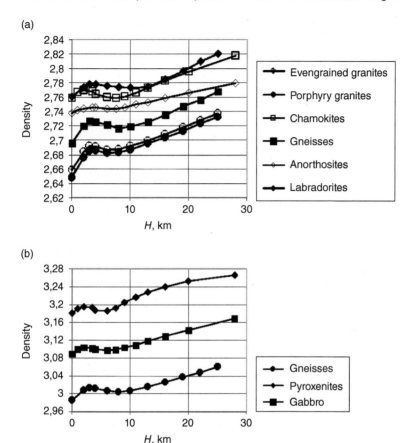

Figure 15.4 Plots of $\rho = f(PT) = f(H)$ for certain rocks.

15.6. Similarity Between Experimental and Observational Data

DSS studies have revealed anomalies in the elastic behavior of mineral material at 3–25 km depths in the so-called LVZs (Figure 15.1). A comparison of interdisciplinary experimental and field data allows us to put forward several ideas on the nature of elastic zoning in the upper layers of the crystalline crust [*Burtny et al.*, 2013; *Korchin*, 2013a; *Korchin et al.*, 2006].

The validity of such a comparison is based on applying relevant similarity criteria that justify a comparative analysis of geophysical observations and laboratory studies of the physical parameters of rocks.

The model and natural processes or phenomena are considered to be similar if the criteria are numerically equal. The theory of dimension and similarity allows us to use effective methods of selecting indispensable relationships [*Korchin*, 2013a]. The criteria of similarity (C) are dimensionless quantities that characterize the given phenomenon or process. A similarity analysis was applied to select the dimensionless quantities. For this purpose, a set of quantities was selected that mostly govern the velocity of elastic waves in mineral material. As the velocity mostly reflects the physical state of matter under different thermal baric conditions, it functionally depends on the pressure (*P*), temperature (*T*), density (ρ), volume (U), specific heat capacity (C_{ou}), and Poisson coefficient (σ):

$$V = f\left(P, T, \rho, U, C_{ou}, \sigma\right). \tag{15.2}$$

Because the researched objects have finite dimensions and are subjected to changing *PT* environments, it is reasonable to consider that *V* depends on *PT* regimes changing with depth. Substituting *T* and *P* by their gradients (*grad T* and *grad P*), we obtain the following dimensions: $[V] = g \cdot s^{-1}$; $[grad\ P] = kg \cdot g^{-2} s^{-2}$; $[grad\ T] = {}^\circ C \cdot m^{-1}$; $[c] = kg \cdot g^{-3}$; $[z] = g$; $[C_{ou}] = m^2 \cdot s^{-2} ({}^\circ C)^{-1}$; $[\sigma]$ is dimensionless.

Solving the matrix of their dimensions, we obtain the following criteria of similarity:

$$C_1 = \frac{gradP \cdot z}{V^2 \cdot \rho} \tag{15.3}$$

$$C_2 = \frac{gradT \cdot z \cdot C_{ou}}{V^2}. \tag{15.4}$$

The first (C_1) characterizes the influence of pressure on the velocity of the elastic wave of the object studied within a boundary thickness of *z*; the second (C_2) is the influence of temperature. Consequently, if a model (*m*) and its natural (*n*) affinity consist of the same matter and $\rho_m = \rho_n$ and $C_{ou\,m} = C_{ou\,n}$, equality of the velocity in the model and natural setting will take place, satisfying the following condition:

$$gradP \cdot z = inv, \ gradT \cdot z = inv. \tag{15.5}$$

Based on these criteria of similarity, the equality of the mean values of *P* and *T* in the model and natural environments is enough to be comparable to the velocity in both these settings. This implies that the V_p/V_S ratio (C_3) must be constant. The mean value of this ratio for the Earth's crust is 1.77, while its laboratory value is 1.8 ± 0.1 [*Korchin et al.*, 2006]. Since the V_p/V_S ratio (C_3) is numerically equal, the velocity of the elastic wave in the relevant model must not differ from that in natural conditions. Thus, the scale coefficients can be taken to be equal to 1. Therefore, equations describing relationships between V_p, X_1, *P*, and *T* from experiments are invariant and can be applied without changes in simulating natural environments at different depths.

The thermobaric petro-velocity modeling is based on the similarity between the elastic velocity in the crystalline crust from DSS data and experiments with samples of different rocks under high temperature and pressures [*Burtny et al.,* 2013; *Korchin,* 2010; *Korchin et al.,* 2013]. The main conclusion of this section is that the LVZs at 3–25 km depth seem to be of thermobaric origin.

15.7. Dynamics of LVZs in the Crystalline Crust

Variations in the thermal regime of the upper crust can result in the change of the LVZs' parameters up to their disappearance [*Korchin,* 2010, 2011a, b, 2012, 2013a, b: *Korchin et al.,* 2013; *Korchin,* 2006]. We experimentally confirmed the strong relationship between the heat conductivity of rocks under different *PT* conditions and their elastic properties. If one supposes that in mineral matter the heat transfer is mainly realized by background heat conductivity, then its value can be estimated using the simple equations of the Debye theory [*Alers,* 1968; *Korchin,* 2011b; *Korchin,* 2013a; *Korchin et al.,* 2013]. As a result of simple transformations, an equation of background heat conductivity was obtained:

$$\lambda_f = \frac{1}{3}\tilde{N}_{ou}\rho V_f l_f = \frac{\delta_0 \beta V_{mid}\rho^{-1/3}}{a_0 \gamma^2 \mu^{2/3}} \cdot \frac{\theta}{T} \approx A \cdot V_m \frac{\theta}{T}, \tag{15.6}$$

where ρ is density; $V_f = V_m$ is average velocity of phonons, which is equal to the average elastic velocity, $(1/V_P^3 + 2/V_S^3)^{-1/3}$; l_f is average length of free motion of phonons; δ_0 is average constant of lattice; β is compressibility; μ is average molecular weight; γ is Grüneisen parameter $\left(\gamma = \dfrac{dLn\theta}{dLn\rho}\right)$; T is temperature; θ is Debye temperature; and A is constant coefficient, including constant parameters independent of *PT* conditions.

This equation implies that the change in heat conductivity in the Earth's crust at some depths is directly proportional to the change in elastic-density characteristics of the mineral matter and inversely proportional to temperature. Based on calculations and experimental studies of $V_p = f(PT), \lambda = f(PT)$ [*Korchin,* 2011b, 2013a], the thermal conductivity of rocks in the Earth's crust changes similarly to $V_p = f(H)$. A plot of $\lambda = f(H)$ demonstrates minimum values coinciding with locations of the LVZs. Thus, the LVZs in the Earth's crust are characterized by lowering values of λ and the reflecting horizon heat flow, whose source is due to deep thermoactive processes.

According to the classic laws of thermodynamics and thermophysics [*Korchin,* 2013b; *Kutas,* 1978; *Nashchekin,* 1969], a layer of lowering heat conductivity during the distribution of thermal energy results in an increase in temperature in lower areas and a decrease in higher ones (Figure 15.5). This process results in the violation of *PT* equilibrium in the LVZ. Decreasing temperature in the upper domain violates the thermobaric condition (1). The state of the rocks in the upper

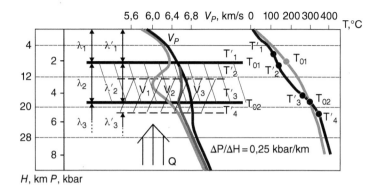

Figure 15.5 Change in parameters of low-velocity zones Vp and deep heat flow (Q).

layer equilibrates with that of the overlying rocks and the upper edge of the zone descends (Figure 15.5). The T_{01} border descends to the T'_2 boundary. At the same time, in the lower part of the underlying rocks, overheating of the zone base occurs proportionally to the difference of heat conductivity of the mineral environment in the zone. This process results in violation of condition (1) and the lower edge of the zone descends to the T'_4 boundary.

Higher pressure stifles the growth of the zone due to the compensating effect of structural distortions under pressure. Thus, the zone modifies its configuration: its thickness can be increased by the increased intensity of deep heat flow, or the zone can disappear in the event of lowering deep heat flow. This instability of thermodynamic LVZs causes their episodic occurrence in the Earth's crust, as well as their vertical and horizontal migration, depending on the temperature variations in the deep horizons of the Earth [*Korchin*, 2013b; *Korchin et al.*, 2013; *Tripolsky and Sharov*, 2004].

The thermobaric petrophysical modeling for the parameters of these zones [*Burtny et al.*, 2013; *Korchin*, 2011a, b; *Korchin*, 2006] (Figure 15.6) has demonstrated that their characteristics depend weakly on the mineral composition of the rocks at relevant depths and are mainly related to the geothermal conditions at similar depths. So at a depth of 25 km, the following dependences of T account for q – heat flow: $T = 250\,°C$ at $q = 30\,mW/m^2$; $T \approx 300\,°C$ at $q = 40$; $T \approx 500\,°C$ at $q = 60\,mW/m^2$. Thus, as the thermal streams in the probed area of the Ukrainian shield vary from 30 to 55 mW/m² (Figure 15.6), different blocks along a type have different temperature gradients with depth and are consequently characterized by different configurations of areas of low seismic velocities. Indeed, apparently in the resulting model (Figure 15.6), LVZs are more significantly and confidently registered in the western and east areas, where $q \approx 50–60\,mW/m^2$, and $T_{25} = 350–425\,°C$ ($\partial T/\partial H \approx 14–17\,°C/km$). In the central block (PK 110–70 km along from DSS, Figure 15.6), where $q \approx 35–45\,mW/m^2$, and $T = 270–305\,°C$, ($\partial T/\partial H < 12\,°C/km$),

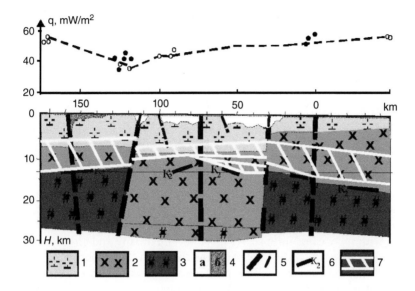

Figure 15.6 Fragment of the schematic cross-section matter composition of the Earth's crust along the IV geotraverse: 1 = plagiogranites; 2 = diorites; 3 = enderbites; 4 = sedimentary-volcanogenic rocks greenstone features (a) and Kirovograd-Zhitomir complex (b); 5 = fault zones; 6 = K_2 boundary; 7 = LVZ; q = heat flow.

the LVZ is less expressed, and lower powered at depth, forming a group in the form of separate lenses. In connection with the lower value of q and the heterogeneity of the thermal field, in the central block (PK 75–30), it is possible to select two areas of low velocity: first, at the depth of H = 6–9 km in granites ($\Delta V \approx 0.15$ km/s) and second, at H ≈ 10–12 km in diorites ($\Delta V = 0.03$ km/s).

A multidisciplinary analysis of data from DSS observations and geothermal and petro-structural modeling has shown that the crystal domains of higher temperature gradients are distinguished by the more complex character of variations in seismic velocity with depth. Here intensive LVZs occur. In "colder" crystal domains, the thickness of the LVZs is negligible or zones are wholly absent. In addition, we confirmed that variations in the mineral composition of rocks at depths of 5–20 km have a negligible influence on the position of LVZs and their intensity.

15.8. LVZs in the Crust as Zones of Increased Porosity of Mineral Matter and Active Transformations of Gas-Liquid Inclusions

The physical parameters of different types of crystalline (igneous and metamorphic) rocks are mainly governed by the composition and structural-textural peculiarities of the mineral material, although the jointing and porosity of rocks and the state of intergrain boundaries also influence these characteristics. The occurrence of different types of pores and fissures is governed by the conditions of their formation and subsequent processes of transformation. Porosity in crystalline rocks has been the subject of intensive studies examining the migration of gas-liquid flows, particularly hydrocarbons, at different depths in the lithosphere [*Geguzin and Krivoglaz*, 1971; *Korchin*, 2012, 2013a, b; *Tripolsky and Sharov*, 2004; *Korchin*, 2006].

It was determined that increasing porosity and microfissures in petrographically similar rocks results in the decreasing velocity of elastic waves. Based on studies of parameters for the sample from the Ukrainian shield under different thermal baric conditions, a change was established in porosity and jointing under different PT regimes. Under atmospheric pressure, a clear relationship was revealed between velocity and porosity for three groups of granites differing in the grain size of the rock-forming minerals. The gradients of changing $V_p/f(n)$ increase with increasing grain size (Figure 15.7). Porosity and structural-textural peculiarities are mean factors that cause the change in the velocity of elastic waves under 3–5 kbar pressure in samples of the same mineral content (Figure 15.7a). The gradients of change in $V_p=f(P)$ are steeper for granites with larger-sized grains in their rock-forming minerals. The velocity of almost poreless granites is in good agreement with extrapolated linear segments of the $V_p=f(P)$ plot to the intersection with the (V_e) ordinate (Figure 15.7b). The curvilinear segments of the $V_p=f(P)$ graphs reflect the increase in the velocity, mainly due to the closing of different types of pores and microfissures (V_{pn}), as well as the change in the elastic properties of the rock-forming minerals ($V_{pn=0}$).

The $V_P = V_0 + (0.6\rho_0 + 0.09n_0 - 1.5)\lg P$ equation was derived from 100 relationships for rocks of different porosity. It quite reliably estimates the V_P of the elastic wave under high pressure (up to 5000 kg/cm²) using the values of the velocity (V_p), density (ρ_0), and porosity (n_0) under atmospheric conditions.

The mean values of pressure under which the pore space does not practically change the velocity are 2.6 (2.0–2.8), 2.2 (1.8–2.6), and 1.9 (1.8–.4) kbar for fine-, medium- and coarse-grained granites, respectively (Figure 15.7c).

The $\dfrac{n_0 - n_P}{n_0} = \dfrac{\Delta n}{n_0} = f(P)$ graphs (Figure 15.7) characterize the relative change in pore space (in percentage) under pressure in the granites, where

$n = n_0 \left(1 - \sqrt{\dfrac{V_P - DP - V_0}{V_e - V_0}}\right)$, and D is a relative change in velocity under pressure on the linear segment due to the elastic deformation of rock-forming minerals. Based on graphical presentations and theoretical calculations, rock porosity decreases by 30%–50% at a depth of 3–5 km (Figure 15.8).

The decreasing velocity of the elastic waves and changing density suggest that in the LVZs at depths of 4–15 km, porosity increases by 10%–20% in comparison with that at 3–5 km depth (horizons of maximum elastic parameters above the LVZ). In the zones of decreased porosity (LVZ), the intensive processes

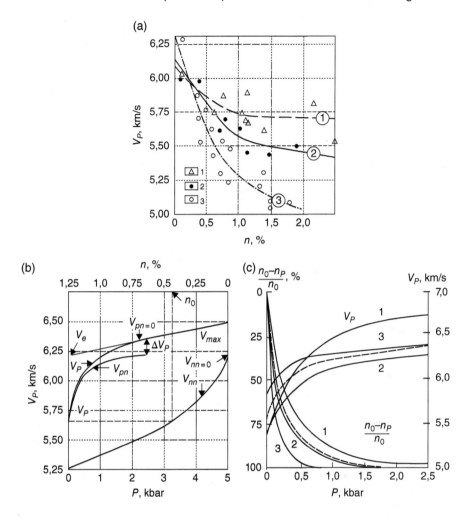

Figure 15.7 Change of $V_p = f(n)$ for granites: 1 = fine-grained, 2 = medium-grained, 3 = coarse-grained (a); change of mean speed values of longitudinal waves for medium-grained granites from pressure ($V_p = f(P)$) and porosity ($V_{Pp} = f(n)$) (b); relative change of pore space under the increase of pressure in granites: 1 = fine-grained, 2 = medium-grained, 3 = coarse-grained (c).

of mass transfer of gas-liquid fluids including deep hydrocarbons seem to be the most intensive [*Korchin*, 2013a, b; *Reider*, 1987]. In light of the data from this paper, the seismic K_2 discontinuity seems to be thermobaric in origin, resulting from the intensive increase in V_p above the LVZ. However, it is not excluded that the intensive increase above the LVZ is additionally associated with the transition

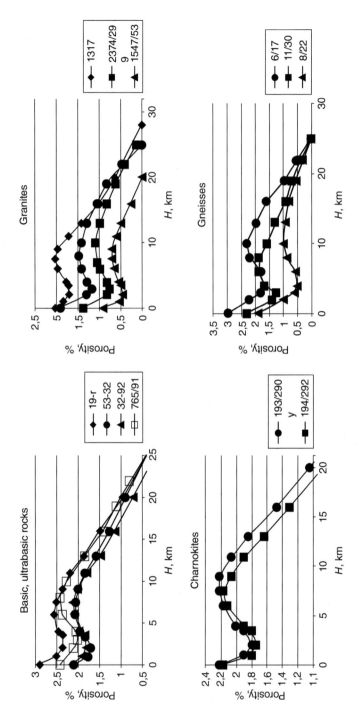

Figure 15.8 Change in porosity of the crystal rocks versus thermodynamic conditions at the relevant depths.

of low-velocity rocks to high-velocity ones. Deep thermobaric conditions facilitate a higher-velocity jump. Among the known deep anomalies of the geophysical fields, LVZs are the most accessible to study using different geological-geophysical methods, including super-deep drilling. This makes it possible to perform the most detailed and promising studies, with the aim of searching for mineral resources and clarifying the deep structure of the Earth, as well as explaining and predicting crustal earthquakes.

15.9. Conclusions

The elastic and structural changes of rocks under thermodynamic conditions at depths of 3–5 km to 12–15 km are characteristic of the properties of rocks during their cataclastic transformations. Within this interval of density lowering of the mineral medium, a dilatancy phenomenon occurs. After gradual existence of the rocks, mechanisms for plastic deformations of the medium arise and improvement of the rock occurs by means of material and structural (at the level of elementary defects) reconstructions. In geological terms, within the interval of depths 20–40 km, the process of regional metamorphism appears to occur. In the inversion zones (3–16 km depth), the multiple correlation coefficients decrease, while outside the LVZ they increase again, although they remain less than in the first area of small depths.

The LVZs in the Earth's crust are considered to be an objective reality resulting from structural transformations of rocks as a consequence of the combined action of pressure and temperature at relevant depths in the lithosphere. They arise when the temperature gradient at relevant depths exceeds a certain threshold and the pressure is not able to compensate the disturbance of the mineral medium produced by temperature. Values of $\partial \rho / \partial H$ are sometimes negative in conformity with the considerable density lowering of rocks that creates zones of low density in the crust. Under the thermodynamic conditions corresponding to 5–15 km depth, the gradient of density increment falls to zero or becomes negative. Additional experiments have revealed that LVZs depend weakly on the mineral composition of the crust matter. Obviously, structural transformations and chemical composition play a key role in the variations in rock density.

In light of the data from this chapter, the seismic K_2 discontinuity seems to be thermobaric in origin, which results from the intensive increase in V_p above the LVZ. However, it is not excluded that the intensive increase above the LVZ is additionally associated with transition of low-velocity rocks to high-velocity ones. Deep thermobaric conditions facilitate a higher-velocity jump.

Among the known deep anomalies of the geophysical fields, LVZs are the most accessible to study using different geological-geophysical methods, including super-deep drilling. It is possible to search for hydrocarbon resources in such zones because rock porosity within them (2.2%–2.78%) fully coincides with that of reservoir rocks (not more than 3%) in Bach Ho (White Tiger) oil field [*Cuong and Warren*, 2009]. In the crystalline basement, the depths of the productive intervals in giant and supergiant petroleum deposits vary from 900 to 5985 m [*Kutcherov and Krayushkin*, 2010]. On the other hand, DSS has registered LVZs at a depth of 3–15 km in the Earth's crust of the Precambrian shields all over the world (Figure 15.1). As the value of depths of hydrocarbon fields and LVZs partly overlap, it follows that LVZs are a promising target to prospect for commercial hydrocarbon accumulations.

References

Alers, D. (1968), Using measurements of sound velocity for determining Debye temperature in solid bodies. In *Dynamics of Lattice*. Mir, Moscow, pp. 13–61 (in Russian).

Anderson, M. P., W. W. Woesserson, and R. J. Hunt (2015), *Applied Groundwater Modeling: Simulation of Flow and Advective Transport*. Academic Press, London.

Burtny, P., V. Korchin, and E. Karnaukhova (2013), *Modeling Material Composition of Deep Horizons of the Crust: A New Conception of the Interpretation of Geophysical Data*. Lambert Academic Publishing, 188 (in Russian).

Cuong, T., and J. K. Warren (2009), Back to field, a fractured granitic basement reservoir, Cuu Long Basin, offshore SE Vietnam: A "buried-hill" play. *Journal of Petroleum Geology*, 32(2), 129–156. doi:10.1111/j.1747-5457.2009.00440.x.

Geguzin, Y., and M. Krivoglaz (1971), *Migration of Macroscopic Inclusions in Solid Bodies*. Metallurgy Publishing, Moscow (in Russian).

Han, J. J., T. H. Lee, and W. M. Sung (2013), Analysis of oil production behavior for the fractured basement reservoir using hybrid discrete fractured network approach. *Advances in Petroleum Exploration and Development* 5(1), 63–70.

Korchin, V. (2006), Thermobaric seismic stratification of the lithosphere. *12th International Symposium on Deep Structure of the Continents and their Margins,* September 24–29, 2006, Shonan Village Centre, Hayama, Japan, NHA-P02, CD-ROM.

Korchin, V. (2010), Thermobaric petrophysical modeling of low velocity zones in the crust. *Physical, Chemical and Petro Physical Studies in Earth Sciences, Materials of International Conference, Moscow*, 147–150 (in Russian).

Korchin, V. (2011a), Petrophysical peculiarities of low velocity zones in the crust and their thermobaric instability. *Materials of the XVII All-Russian Conference with International Participation: "Problems of Seism Tectonics,"* Moscow, 20–22 September 2011, pp. 273–277 (in Russian).

Korchin, V. (2011b), Influence of thermal regimes on crustal low velocity zones. *Proceedings of International Conference: "6th Yu. P. Bulashevitch Reading", Ekaterinburg, Russia*, pp. 199–201 (in Russian).

Korchin, V. (2012), Elastic-density thermobaric stratification of the lithosphere, crustal low velocity zones. *Materials of III Tectonophysical Conference in IPE RAS: Tectonophysics and Current Problems of Earth Sciences, Moscow*, 8–12 October 2012, vol. 2, pp. 390–393 (in Russian).

Korchin, V. (2013a), *Thermobaric of Crustal Low-Velocity Zones (A New Scientific Hypothesis)*. Lambert Academic Publishing, 280 (in Russian).

Korchin, V. (2013b), Crust low velocity zones: Perspective horizons for localization of deep hydrocarbons. *Deep Oil*, 8, 1099–1116 (in Russian).

Korchin, V., P. Burtny, and E. Karnaukhova (2006), Thermobaric petro structural modeling of the Earth's crust and the nature of some seismic boundaries. *12th International Symposium on Deep Structure of the Continents and Their Margins*, 24–29 September 2006, Shonan.

Korchin, V., P. Burtny, and V. Kobolev (2013), *Thermobaric Petrophysical Modeling in Geophysics*. Kiev, Naukova Dumka (in Russian).

Kun, W., and H. Suyun (2015), Two quantitative evaluation methods for identifying the migration pathways in hydrocarbon carrier: Application and comparison. *The Open Petroleum Engineering Journal*, 8, 172–180.

Kutas, R. (1978), *Field of Heat Flows and a Theoretical Model of the Earth's Crust*. Naukova Dumka, Kiev (in Russian).

Kutcherov, V. G., and V. A. Krayushkin (2010), A deep-seated biogenic origin of petroleum: From geological assessment to physical theory. *Reviews of Geophysics*, 48. doi:10.1029/2008RG000270.issn:8755-1209.

Nashchekin, V. (1969), *Technical Thermodynamics and Heat Transfer*. Higher School, Moscow (in Russian).

Reider, E. (1987), *Fluid Inclusions in Minerals*. Mir, Moscow (in Russian).

Sircar A. (2004), Hydrocarbon production from fractured basement formation. *Current. Sci.*, 2, 147–151.

Tripolsky, A., and N. Sharov (2004), *Lithosphere of the Precambrian Shields of the Northern Hemisphere as Derived from Seismic Data*. Karelian Scientific Centre RAS, Petrozavodsk (in Russian).

16

16. THE USE OF AEROMAGNETICS AND MICROMAGNETICS TO IDENTIFY POTENTIAL AREAS OF HYDROCARBONS IN THE MIDCONTINENTAL UNITED STATES: CAVEATS AND PITFALLS

Steven A. Tedesco

Abstract

Aeromagnetics and its derivative micromagnetics represents an exploration tool for petroleum exploration. Micromagnetics anomalies are typically, but not always caused by the presence of migrating hydrocarbons from depth that react in the soil or near surface and cause precipitation of magnetite minerals. Presented here is a detailed aeromagnetic survey located in Anderson and Linn counties in the southern portion of the Forest City Basin, Kansas, Mid-Continent USA. The survey is in an optimal area of thin Paleozoic cover of less than 4,000 feet and the target reservoirs are less than 1,500 feet. Aeromagnetics is well documented in its ability to define fault systems in the Precambrian basement where the overburden strata are relatively thin. The survey found several areas of micromagnetic anomalies that do not seem to be related to man-made materials, existing oil fields, soil and bedrock changes. There are several areas of micromagnetic anomalies that are related to several existing producing fields. Micromagnetics provide an inexpensive approach to targeting and prioritizing areas for hydrocarbon exploration. These areas can be followed up with other methods such as surface geochemistry, gravity, electrical and seismic prior to drilling.

Atoka Inc., Englewood, Colorado, USA

Oil and Gas Exploration: Methods and Application, Monograph Number 72,
First Edition. Edited by Said Gaci and Olga Hachay.
© 2017 American Geophysical Union. Published 2017 by John Wiley & Sons, Inc.

16.1. Introduction

The use of aeromagnetics in petroleum exploration to define basement structural features has been well established [*Gay*, 1996]. The use of micromagnetics is also been useful, but its results are less definitive. The presence of magnetic minerals in certain cases in the near surface has been related to seeping hydrocarbons that form under a variety of soil conditions. The literature has several case histories that show strong correlation between micromagnetic anomalies and existing oil fields. A survey was flown over portions of Anderson and Linn Counties, Kansas, Forest City Basin, Midcontinent USA. The aeromagnetic survey defined basement features and in many cases indicated fault systems that strongly coincide with existing petroleum fields. The results of two independent micromagnetic analyses presented different results.

The use of micromagnetics has been well established [*Donovan* 1974; *Donovan et al.*, 1979; *Elmore*, 1990; *Tedesco*, 1994; *Berger et al.*, 2002; *Wollenben and Greenlee*, 2002; *LeShack and Alstine*, 2002] for finding petroleum reservoirs. The difficulty in using these methods is screening out anomalous areas that are not related to petroleum reservoirs. The cause of these shallow or near-surface magnetic mineral-rich areas is the presence of hydrocarbons that have seeped up from the subsurface, presumably a petroleum filled reservoir, initiating the precipitation of magnetic minerals. These magnetic minerals are deducted during the normal course of either a ground or airborne magnetic survey. By screening or removing out the magnetic data related to the basement, the remaining data is related to near-surface events, man-made, geologic, or otherwise. These data then are evaluated to determine what is not being caused by soil, bedrock, and man-made material and is potentially related to seeping hydrocarbons from depth.

16.2. Geology

The study area located in the Forest City Basin is a shallow, asymmetric, intracratonic basin that is bounded to the west by the Nemaha Ridge, to the south by the Bourbon Arch, to southeast by the Ozark Uplift, to the east by the Lincoln Anticline, and to the north by the Precambrian Shield (Figure 16.1). The basin formed post-Devonian time as a result of movement along the Nemaha Ridge and the breakup of the early Paleozoic age Kansas Basin [*Tedesco*, 2014].

The Precambrian rocks (Figure 16.2) underlying the survey area are a wide band of thick, alternating hard and soft layered clastics that are only slightly metamorphosed that trend northwestward across northeastern Kansas and constitute a regional province [*Merriam*, 1963]. The Precambrian is overlain by the Reagan sandstone of the Cambrian age (Figure 16.2), which varies from a few ft to several hundred ft as it filled in topographic lows on the Precambrian surface.

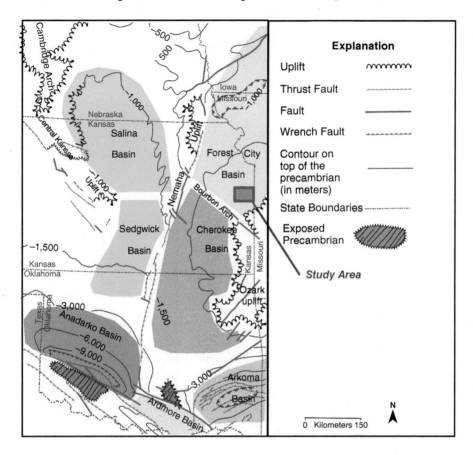

Figure 16.1 Location of the aeromagnetic survey in Anderson and Linn Counties, Forest City Basin, Kansas, USA.

The Arbuckle Group of late Cambrian to early Ordovician overlies the Reagan sandstone (Figure 16.2) and consists of fractured limestones and dolomites with thin interbedded shale and sandstones and is 750 to 950 ft (227 to 287 m) thick [*Merriam*, 1963]. The sandstones and shale decrease upward in the group. The Arbuckle Group is overlain by the middle Ordovician age Simpson or St. Peter sandstone, which is an orthoquartzarenite that is only present in the northwest part of the study area and is 0 to 25 ft (0 to 7 m) thick [*Merriam*, 1963]. The formation of the Chautauqua Arch in Silurian and Devonian times eroded off all prelate Devonian to as old as early Ordovician rocks in the study area [*Merriam*, 1963; *Tedesco*, 2014].

The late Devonian to early Mississippian Chattanooga Shale overlies the Arbuckle Group in the study area (Figure 16.2). The Chattanooga is a highly

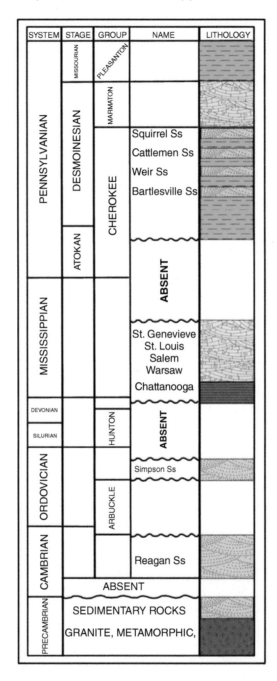

Figure 16.2 Stratigraphic column for the study area in Anderson and Linn Counties, Kansas, southern Forest City Basin, Midcontinent USA.

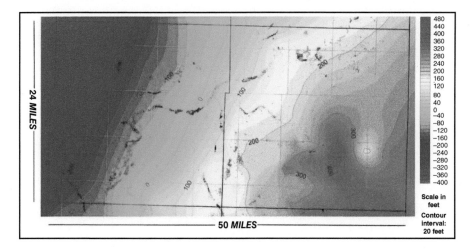

24 MILES

50 MILES

480
440
400
360
320
280
240
200
160
120
80
40
0
−40
−80
−120
−160
−200
−240
−280
−320
−360
−400

Scale in feet

Contour interval: 20 feet

Figure 16.3 Structure on top of the Mississippian carbonates in the study area.

organic shale (2% to 12% total organic carbon), 20 to 50 ft (5 to 15 m) thick, that is the potential source of petroleum for hydrocarbons trapped in reservoirs in the area. Overlying the Chattanooga rocks are the Mississippian carbonates (Figure 16.2) that are primarily porous or dense limestones whose top surface is eroded, karsted and contains significant chert [*Merriam*, 1963]. The Mississippian varies from 280 to 400 ft (85 to 120 m) thick. Structure on top of the Mississippian carbonates is indicated in Figure 16.3. Petroleum test wells are represented by black symbols (dry holes) and green dots (oil producers). The Mississippian surface structural strike is to the northwest-southeast. Petroleum production in the area is less than 1,500 ft (450 m), and the Mississippian carbonates or older rocks to date have not been proven prospective for petroleum production; therefore, the majority of the drilling does not reach this rock unit.

The Mississippian rocks are overlain by the Pennsylvanian age Cherokee Group (Figure 16.2) that is predominantly lacustrine to near-marine gray-green shales interbedded with thin coals, limestones, carbonaceous mudstones, and sandstones deposits that are deltaic, tidal, and fluvial deposited across a coastal plain [*Merriam*, 1963; *Tedesco*, 2014]. The Cherokee Group varies from 300 to 440 ft (90 to 133 m) thick [*Tedesco*, 2014]. Within the Cherokee Group is found the oil productive Squirrel, Cattleman, Weir, and Bartlesville sandstones that vary from 15% to 26%, 3 to 50 ft (1 to 15 m) thick, 10 md to over a Darcy permeability at less than 1,500 ft in the study area. The basin is deepening from east to west. Figure 16.4 is the structure on top of the Cherokee Group, and as the petroleum productive reservoirs are below this marker, it gives a more definitive indication that the structural strike is northwest-southeast. The isopach of the Cherokee Group is indicated in Figure 16.5. The Cherokee Group thickens toward the

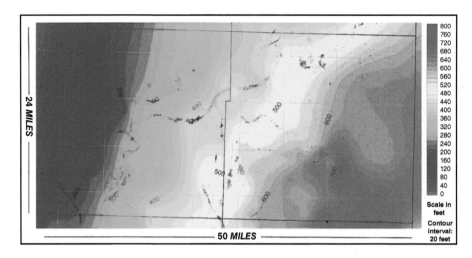

Figure 16.4 Structure on top of the Cherokee Group in the study area.

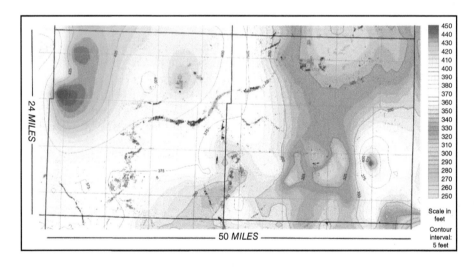

Figure 16.5 Isopach of the Cherokee Group in the study area.

northwestern corner of the map, which is toward the basin depocenter. There is generally no relationship between the thickening of the Cherokee Group or Cherokee and Mississippian structure.

The Marmaton Group overlies the Cherokee Group and in most of the study area is at the surface and the youngest rock unit present. The Marmaton consists predominantly of cyclic marine limestones with thin carbonaceous mudstone, nonorganic shale, coal, and minor marine or near-shore sandstones. The Group varies from 0 to 100 ft thick. Overlying the Marmaton in the western part of the study area is the Pleasanton Group (Figure 16.2). The group is primarily noncarbonaceous shale and is less than 100 ft thick.

Petroleum production in the area is from Cherokee Group fluvial, tidal, and near-shore sandstones that have meandered across the study area. Some of the productive sandstones are part of a deltaic system. The poor quality of data makes it difficult to accurately predict any particular fluvial trend outside the existing well data. Because of the shallow nature of the productive sandstones (<1,500 ft, 450 m) the use of seismic is cost prohibitive. Exploration is based essentially on subsurface mapping geology and random drilling.

16.3. Aeromagnetics

An aeromagnetic survey was acquired by CGG for Running Foxes Petroleum Inc. over the study area in 2013 at 152 m, 200-m lines with tie-lines at 800 m. The direction of the lines was 90° and tie lines 0° with a total of 13,002 km of data acquired. The dominant flight pathways were west to east, which is oblique to known structural trends.

The aeromagnetics were processed to determine the magnetic characteristics of the basement in the study area by using standard geophysical processing methods [*Grant and West*, 1965; *Hartman et al.*, 1971; *Nettleton*, 1976; *Sharma*, 1976; *Dobrin*, 1976]. One of those methods is the reduction-to-the-pole (RTP) transformation operation, a data processing technique that recalculates total magnetic intensity data as if the inducing magnetic field had a 90° inclination. This transforms dipolar magnetic anomalies to monopolar anomalies centered over their causative bodies, which can simplify the interpretation of the data. RTP makes the simplifying assumption that the rocks in the survey area are all magnetized parallel to the earth's magnetic field. This is true in the case of rocks with an induced magnetization only. However, remanent magnetization will not be correctly dealt with if the direction of remanence is different to the direction of the earth's magnetic field. In sedimentary basins, remanence is usually not a problem. In the majority of cases, the RTP transform is stable. RTP does not work as well close to the magnetic equator (<10° declination) as there is a large correction to be made for the amplitude of the anomalies. This is usually addressed using specially designed variations of the RTP transform. Errors in the RTP transform usually appear as narrow anomalies elongated parallel to the declination of the earth's magnetic field. The RTP transform can be applied to either gridded or profile data but is usually only applied to grids.

The RTP technique essentially removes the Earth's magnetic inclination and declination effects on the measured magnetic data, and simply shifts positive magnetic anomalies directly over areas of positive magnetic contrasts and

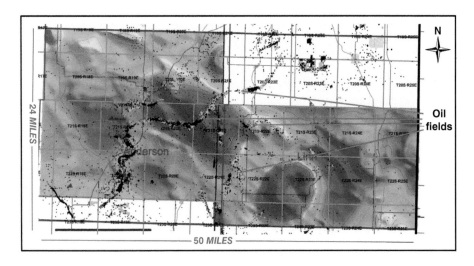

Figure 16.6 Reduction to pole map.

negative anomalies over negative magnetic contrasts. This vastly simplifies the interpretation of magnetic anomalies and makes them more understandable and intuitive, like gravity anomalies. As is demonstrated in Figure 16.6, a total magnetic intensity anomaly (as would be measured in an airborne magnetic survey), is not symmetric to the causative body within the Earth. The RTP technique uses the geographic location of each point of data to calculate the Earth's inclination, declination, and field strength and uses that information to shift the magnetic anomaly directly over the causative body. This is what the magnetic data would look like if the data were collected at either the north or south magnetic pole, hence the name "reduction to the pole."

The data were interpreted by both CGG and Earthfield Technologies. The RTP magnetic map for the study area is presented in Figure 16.6 and indicates a northwest-southeast trending feature that is consistent with the structural trend in the area. The study has been interpreted by previous work and authors to have a series of intrusives of either Precambrian age or potentially as young as Permian [*Yarger*, 1983]. Internally, there are several minor features within the larger feature that have a similar trend. The RTP residual indicates a similar trend (Figure 16.7). Both figures show a northwest-southeast structural grain that is consistent with the known structure fabric in the area.

The basement fault maps are produced using the Werner deconvolution method of quantitative magnetic analysis [*Ku and Sharp*, 1983, 1984; *Rao*, 1984]. The Werner technique is a profile based, inverse modeling method that calculates a series of depth estimates to the various magnetic anomalies that have been measured. These profiles are then analyzed by the geophysicist to interpret

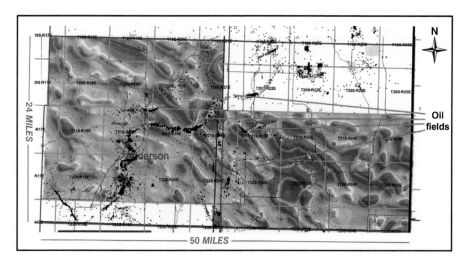

Figure 16.7 Reduction to pole map with a 7.5 km pass.residual filter.

possible fault locations at the basement surface. After all of the magnetic flight profiles have been analyzed, and faults mapped in two dimensions, the results are combined to produce a fault trace map (basement fault map).

Werner deconvolution theory describes magnetic anomalies that can be produced by a number of causative features such as lithology changes, variations in the thickness of magnetic units, faulting, folding, and topographic relief. A significant amount of information can, at times, be obtained from a qualitative review of the magnetic anomaly maps. Magnetic anomaly wave numbers can often be related to the depth of anomalous sources from the magnetic sensor. Given the same rock type (equivalent susceptibilities), shallow sources produce higher wave number anomalies while deeper sources yield broader anomalies. At the same time, variations in the relative magnetic base level can be indicative of regional or subregional variations in basement lithology. Also, the analysis of properly designed residual/transform maps in areas of known or suspected structural tendencies can be of great value in recognizing and mapping more subtle targeted structures.

Magnetic data can be analyzed in a number of ways by computer to yield source body parameters quickly and accurately. The primary method used in this interpretation utilized Werner deconvolution–based algorithms [*Werner*, 1953] as a basis for the interactive analyses of the depth to and horizontal position of source bodies and the related parameters of dip and susceptibility. The method forms a rigorous, iterative, two-dimensional inversion technique that eliminates interference from adjoining, often overprinted, anomalies. The PC-based program MAGDEPTH was utilized in this study for Werner deconvolution.

Analysis of the total magnetic intensity data yields parameters for thin, sheet-like bodies such as dikes, sills, intruded fault zones, and basement plates of minor relief compared to the source-sensor separation distance. Analysis of the horizontal gradient data (computed each pass from the filtered total intensity) yields similar parameters for geologic interface type features such as dipping contacts, edges of prismatic bodies, major faults, and slope changes of the basement surface. All thin sheets are bounded by two interfaces, but these are too close together for their individual anomalies to be distinguishable from one another as the combination of the two results in a single thin sheet anomaly. It has often been shown that the magnetic anomaly for a thin sheet is precisely the same as the horizontal derivative of the magnetic anomaly for a similarly positioned interface. Thus, if interface-type anomalies are present in total field data, they can be transformed to thin sheet–type anomalies simply by calculating the derivative of the total field data. The similarity between the shapes of total field and horizontal derivative anomalies for thin sheets and interfaces is illustrated in Figure 16.1.

The power of the equations employed for direct analysis of magnetic anomalies permits a unique solution even when only parts of the anomaly are undisturbed by adjacent anomalies. Such interference effects are very often the rule rather than the exception. Solutions are found from any set of equally spaced data points, and these points are not confined to critical points along the profile such as minima, maxima, or changes in slopes.

Sample spacing between adjacent data points forming the deconvolution operator, the order and terms of the filter to be used, and the order of the interference operator to be applied to the data are all selected by the interpreter for each of the six operator passes in order to extract the maximum amount of information possible for any given geologic province and structural environment. The horizontal distance subtended by the observational array (the operator length) is critical to the detection of an anomalous source. This distance must be increased to insure that progressively deeper geologic sources having broader magnetic anomalies may be recognized. As such, the computation sequence described above can be performed for up to six passes, allowing for increasing operator sizes each time. The operator length for each pass is established through the setting of the spacing parameter for each pass that specifies data decimation. There often can be sufficient overlap between the depths ranges covered by each pass such that an anomaly is "recognized" on multiple passes.

The spacing parameters for each operator pass used in this interpretation were based on a 328-ft (100 m) sample spacing. The actual operator length is a function of the data spacing, as acquired, the subsample interval used, and the order of interference chosen for each pass (first, second, or third corresponding to a six-, seven-, or eight-point operator). During the deconvolution, the six, seven, or eight equally spaced points along the profile are used to solve the set of simultaneous equations yielding, if an anomaly has been recognized, depth and along line position of the related source body. The entire sequence of points (the

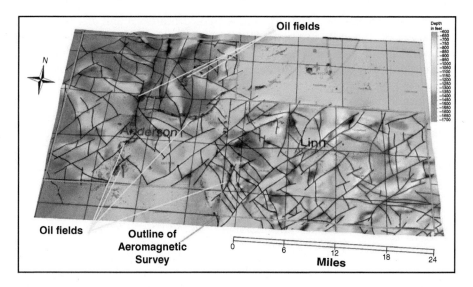

Figure 16.8 Basement fault interpretation in the study area with oil and gas fields. Contour interval in feet from sea elevation. The oil (green) and gas (red) wells are plotted on the map along with dry holes (black).

operator) is then advanced by one point (regardless of the actual operator spacing and length) and another calculation is made. In this way, as many X and Z pairs are obtained as there are data points along an anomaly. The calculated positional coordinates are then examined statistically for consistency, and inconsistent depth calculations are rejected. If the operator is passing over an anomaly, there will be a closely grouped set of depth points indicating a source for the anomaly, and a central solution position can be determined. Susceptibility and dip values are then computed.

When applied to adequately sensitive data, this technique can identify and resolve very subtle anomalies and yield a significant amount of information. In areas where the Werner deconvolution analysis failed to yield depth solutions, manual techniques, such as half maximum slope, were attempted to establish source depths. It is then for the interpreter to analyze the total set of source depths, accept or reject solutions based on their consistency, and apply any known geologic constraints to the interpretation.

The data were analyzed for basement faults as defined by aeromagnetics. Figure 16.8 is the basement fault interpretation that clearly indicates a close association to many of the petroleum productive areas.

The basement fault interpretation is presented in Figure 16.8 along with the existing oil and gas wells. *Gay* [1996] established that many petroleum productive reservoirs are related to magnetic variations such as basement faults. These fault systems can provide pathways for migrating hydrocarbons and are part of the formation of the trap for petroleum in the reservoir. In Figure 16.8 blue and purple colored areas represent structural lows on the Precambrian basement. The faulting trends mimic the same northwest-southeast trend as Figures 16.6 and 16.7. There are numerous cross-faults, causing offsetting basement blocks to have a stairstep pattern, which indicates a wrench and strike-slip fault system. The producing oil fields are strongly associated with several basement faults of areas at the juncture between where faults occur. This suggests the sandstone channels deposition was controlled by basement faulting. These fault systems would have extended into the Paleozoic sedimentary section and would have acted as pathways for migrating petroleum [*Tedesco*, 2014].

16.4. Micromagnetics

Micromagnetic analysis involves the use of high-resolution, airborne, magnetic data, collected along a very tight data grid. The data often contain numerous (and localized), high-frequency magnetic anomalies either associated with shallow, sedimentary-based sources or caused by microseepage of hydrocarbons from a petroleum accumulation at depth into the soil horizon [*Mullins and Tite*, 1973; *Donovan*, 1974; *Donovan et al.*, 1979; *Elmore and Crawford*, 1990; *Reynolds et al.*, 1990; *Reynolds et al.*, 1991; *Newell et al.*, 1993; *Tedesco*, 1994; *Berger et al.*, 2002; *Wollaben and Greenlee*, 2002; *LeShack and Alstine*, 2002]. Figure 16.9 is a typical profile whereby the micromagnetic anomalies are isolated. The seeping hydrocarbons

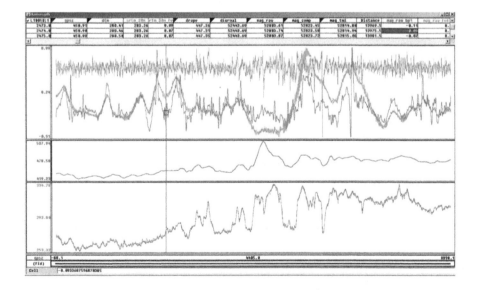

Figure 16.9 Isolation of micromagnetic anomalies using data processing.

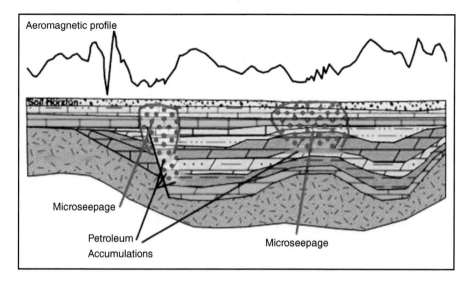

Figure 16.10 Concept behind the formation of micromagnetic anomalies as based on hydrocarbon microseepage.

(Figure 16.10) create a reducing environment that allows magnetic minerals (like magnetite) to precipitate and collect in a vertical "chimney" above the hydrocarbons [*Machel and Burton*, 1991a, b]. The accumulation of the magnetic minerals in the shallow section can then be identified (under ideal conditions) and delineated with a high-resolution, magnetic survey. The caveat to magnetic minerals being found in the soil or near surface may be the result of localized soil or bedrock changes and not related to migrating hydrocarbons.

The analytical approach is to delineate high-frequency anomalies of interest, from deeper-sourced anomalies arising from basement, intrabasement, or surface sources potentially caused by microseepage. Frequency filtering of the magnetic data is done in both grid and profile form. The profile-based filtering is the most effective in removing the isolation of the possible "hydrocarbon chimney"–related anomalies. Progressively higher-frequency, high-pass filters are used until the anomalous areas are determined to be optimal.

The anomalies from the profile-based filtering are then compared to digital well locations and cultural information, since high-frequency anomalies can be produced by these sources as well. Anomalies not related to culture, wells, or pipelines are then identified and mapped. The resultant map displays a series circular to elongated polygons or areas that identify localized concentrations of high-frequency anomalies that do not appear to be culturally related. These sources are not neccesarily related to hydrocarbon accumulations but can possibly be. The anomalous ares can be related to soil or bedrock variations, and these have to be analyzed and corrected for.

16.4.1. Evaluation of the Micromagnetics

Two different analyses and interpretations of micromagnetics in the study area were done. The area is dominated by farming and livestock with towns of generally fewer than 1,000 people. The terrain is relatively flat with small streams and tributaries that have less than 100 ft (30 m) of relief. The first interpretation indicates the micromagnetic anomalous areas as orange polygons through the use of a high-grade filter (Figure 16.11). The existing fields can be clearly seen with micromagnetic anomalies due to the presence of production pipe and surface equipment. There are very linear anomalous trends that are represented as related to gas or oil pipelines. Other linear anomalies coincide with interpreted basement faults as indicated in Figure 16.8. There is no specific trend that implies any undiscovered fields related to Cherokee Group fluvial sandstone trend. The existing fields and pipelines are clearly identified, but the micromagnetic anomalies are probably the result of steel casing present and not magnetic minerals formed from seeping hydrocarbons. The anomalies may represent leakage from deeper petroleum reservoirs or fault systems not identified by the aeromagnetic survey. Similar anomaly interpretation problems were encountered by *Donovan* [1974], *Donovan et al.* [1979] *Tedesco* [1994], *Berger et al.* [2002], and *LeShack and Alstine* [2002].

The second interpretation is presented in Figure 16.12. These are much larger areas. The linear anomalies associated with the existing pipelines, roads, towns, and oil fields are removed. Several faults are interpreted that coincide with similar faults as indicated Figure 16.8. Some of the anomalies coincide with some of the existing fields. In Linn County, several of the anomalies have a

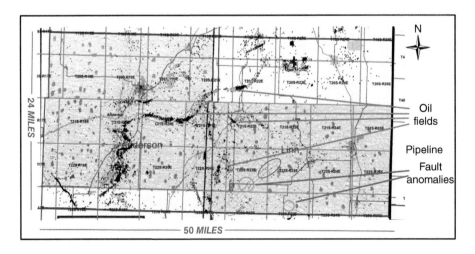

Figure 16.11 Interpretation of micromagnetics in the study area using a high-grade filter.

northwest-southeast feature coinciding with the structural trend as seen in Figures 16.3 and 16.4.

16.5. Conclusion

The use of micromagnetics represents an exploration tool for petroleum exploration. The use of the micromagnetics represents the presence of microseepage from migrating hydrocarbons that react in the soil or near surface and cause precipitation of magnetite minerals. Other causes of the anomalous areas can be changes in soil, bedrock, presence of man-made materials, or causes unknown.

The study presented here in Anderson and Linn Counties in the southern portion of the Forest City Basin represents an optimal area because the Paleozoic cover is thin, less than 4,000 ft, and the petroleum reservoirs are at a shallow depth, less than 1,500 ft, which allows for minimal distance of microseepage to the surface. Aeromagnetics can more accurately identify fault systems in the Precambrian basement because of the thin Paleozoic cover. The numerous anomalies identified in both Figures 16.11 and 16.12 represent those areas of interest that do not seem to be related to man-made materials, existing oil fields, and soil and bedrock changes. Each anomalous area needs to be evaluated by using other surface geochemical methods, such as soil gas, that directly detect hydrocarbons. Additional methods such as subsurface geologic mapping, drilling, or seismic are needed to further define whether these anomalies are real or false. Micromagnetics represent a relatively inexpensive tool that can assist in petroleum exploration.

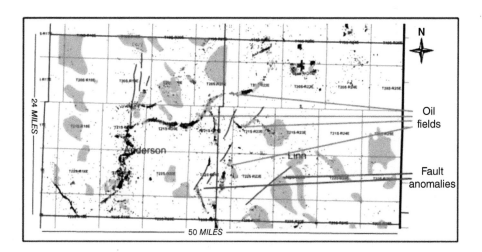

Figure 16.12 Interpretation of micromagnetics in the study area using a less-than-discriminating filter.

References

Berger, Z., J. Davies, and R. T. Thompson (2002), Integrated analysis of high-resolution aeromagnetic (HRAM) and RADARSAT-1 imagery for exploration in mature and frontier basins. In D. Schumacher and L. A. LeSchack (eds.), *Surface Case Histories, Application of Geochemistry, Magnetics and Remote Sensing*, AAPG Studies in Geology No. 48, American Association of Petroleum Geologists.

Dobrin, M. B. (1976), *Introduction to Geophysical Prospecting*, 3rd ed. McGraw-Hill, New York.

Donovan, T. J. (1974), Petroleum microseepage at Cement, Oklahoma: Evidence and mechanism. *Am. Assoc. Petr. Geol. Bull.*, 58(3), 429–446.

Donovan, T. J., R. L. Forgey, and A. A. Roberts (1979), Aeromagnetic detection of diagenetic magnetite over oil fields. *Am. Assoc. Petr. Geol. Bull.*, 63(2), 245–248.

Elmore, R. D., and L. Crawford, L. (1990), Remanence in authigenic magnetite: Testing the hydrocarbon magnetite hypothesis. *Journal of Geophysical Research*, 95, 4539–4549.

Gay, S. P. (1996), Selected basement control of select oil and gas fields in Kansas as determined by detailed residual aeromagnetic data. Select Geophysical Oil and Gas Fields in Kansas, KGS Bulletin 237, 10–16.

Grant, F. S., and G. F. West, G. F. (1965), *Interpretation Theory in Applied Geophysics*. McGraw-Hill, New York.

Hartman, R. R., D. J. Tesky, and J. L. Friedberg (1971), A system for rapid digital aeromagnetic interpretation. *Geophysics*, 36, 891–918.

Ku, C. C., and J. A. Sharp (1983), Werner deconvolution for automated magnetic interpretation and its refinement using Marquardt's inverse modeling. *Geophysics*, 48(6), 754–774.

Ku, C. C., and J. A. Sharp (1984), Reply to discussion of "Werner deconvolution for automated magnetic interpretation and its refinement using Marquardt's inverse modeling." *Geophysics*, 49, 1119.

LeShack, L. A., and D. R. Van Alstine (2002), High resolution ground-magnetic (HRGM) and radiometric surveys for hydrocarbon exploration: Six case histories in western Canada. In D. Schumacher and L. A. LeSchack (eds.), *Surface Case Histories, Application of Geochemistry, Magnetics and Remote Sensing*, AAPG Studies in Geology No. 48, American Association of Petroleum Geologists.

Machel, H. G., and E. A. Burton (1991a), Causes and spatial distribution of anomalous magnetization in hydrocarbon seepage environments. *Am. Assoc. Petr. Geol. Bull.*, 75(12), 1864–1876.

Machel, H. G., and E. A. Burton (1991b), Chemical and microbial processes causing anomalous magnetization in environments affected by hydrocarbon seepage. *Geophysics*, 56(5), 598–605.

Merriam, D. F. (1963), The geologic history of Kansas, *Kansas Geological Survey Bulletin*, 162, 317p.

Mullins, C. E., and M. S. Tite (1973), Magnetic viscosity, quadrature susceptibility, and frequency dependence of susceptibility in single-domain assemblies of magnetite and maghemite. *J. Geophysical. Research*, 78(5), 804–809.

Nettleton, L. L. (1976), *Gravity and Magnetics in Oil Prospecting*. McGraw-Hill, New York.

Newell, A. J., D. J. Dunlop, and W. Williams (1993), A two-dimensional micromagnetic model of magnetizations and fields in magnetite. *J. Geophysical. Research*, 98(B6), 9533–9549.

Rao, V. B. (1984), Discussion of "Werner deconvolution for automated magnetic interpretation and its refinement using Marquardt's inverse modeling" by C. C. Ku and J. A. Sharp. *Geophysics*, 49, 1119.

Reynolds, R. L., N. S. Fishman, and M. R. Hudson (1991), Sources of aeromagnetic anomalies over Cement oil field (Oklahoma), Simpson oil field (Alaska), and the Wyoming-Idaho-Utah thrust belt. *Geophysics*, 56(5), 606–617.

Reynolds, R. L., M. Webring, V. J. S. Grauch, and M. Tuttle (1990), Magnetic forward models of Cement oil field, Oklahoma, based on rock magnetic, geochemical, and petrologic constraints. *Geophysics*, 55(3), 344–353.

Sharma, P. V. (1976), *Geophysical Methods in Geology*. Elsevier Science, New York.

Tedesco, S. A. (1994), Surface geochemistry in petroleum exploration, Chapman & Hall, New York.

Tedesco, S. A., (2014), *Reservoir characterization and geology of the coals and carbonaceous shales of the Cherokee Group in the Cherokee Basin, Kansas, Missouri and Oklahoma USA*, unpublished PhD dissertation, Colorado School of Mines, 2,023p.

Werner, S. (1953), Interpretation of magnetic anomalies at sheet like bodies: *Sveriges Geol. Undersok.*, Series C, Arsbok. 43 (1949), no. 6.

Wollaben, J. A., and D. W. Greenlee (2002), Successful application of micromagnetic data to focus hydrocarbon exploration. In D. Schumacher and L. A. LeSchack (eds.), *Surface Case Histories, Application of Geochemistry, Magnetics and Remote Sensing*, AAPG Studies in Geology No. 48, American Association of Petroleum Geologists, Tulsa, Oklahoma, pp. 67–156.

Yarger, H. L. (1983), Regional interpretation of Kansas aeromagnetic data. *Kansas Geological Survey Geophysics Series*, 1.

INDEX

Oil and Gas Exploration: Methods and Application, Monograph Number 72,
First Edition. Edited by Said Gaci and Olga Hachay.
© 2017 American Geophysical Union. Published 2017 by John Wiley & Sons, Inc.